GIS Mapping and Its Applications in Water Resources Management

NIPA® GENX ELECTRONIC RESOURCES & SOLUTIONS P. LTD.
New Delhi-110 034

About the Editors

Dr. Kommireddi Chinni V Naga Kumar, is working as a 'Scientist' at Land and Water Management Research Group, KSCSTE- Centre for Water Resources Development and Management (CWRDM). He completed his PhD in Geo-Engineering (Remote Sensing and GIS) and M.Tech in Remote Sensing and GIS. He is a recipient of "Prof. K. Nageswara Rao Endowment Young Achiever Award-2016" given by the Indian Society of Geomatics (ISG) and International Union for Quaternary Research (INQUA) Travel grant under Young Researcher category for attending INQUA Congress held at Nagoya Japan during 26 July–2 August, 2015. He is a member of the Indian Society of Remote Sensing and the Indian Society of Geomatics. His research interest includes Coastal Morphodynamics and Holocene Delta Evolution. He has more than 40 publications and is handling several research projects pertaining to "Water Resources" funded from National & International agencies..

Dr. Naveena K, Scientist, LWMRG, CWRDM completed his PhD in Agricultural Statistics. He is a recipient of the Young Achiever Award, Society for Advancement of Human and Nature (2015), SP Dhall Distinguished Publication Award in Statistics Society for Advancement of Human and Nature (2015), and the Young Scientist Award ASTHA Foundation 2015. He is a member of the Indian Society of Agricultural Statistics (ISAS), Statistics Society for Advancement of Human and Nature (ASTHA) Foundation, and the Science and Tech Society for Integrated Rural Improvement. His research interests include Machine Learning Modeling, Forecasting Modeling, Multivariate Analysis, and Design of Experiment. He has more than 30 national and international publications.

Dr. Surendran U, Principal Scientist, Division of Remote Sensing Application, ICAR-National Bureau of Soil Survey and Land Use Planning, Nagpur, completed his PhD and MSc in Soil Science and Agricultural Chemistry. He is a recipient of the Young Achiever award from SADHNA, Young scientist project by the Department of Science and Technology (DST-SERC-Fast track). He is the editorial board member of Agricultural Water Management, Elsevier publishers, handling editor of Scientific Reports, nature publishing group, and member of the International Society for Agricultural Meteorologists, Netherlands, etc. His research interests include

NIPA® GENX ELECTRONIC RESOURCES & SOLUTIONS P. LTD.

101,103, Vikas Surya Plaza, CU Block
L.S.C. Market, Pitam Pura, New Delhi-110 034
Ph : +91-11-43860225, Mob.: +91 9717133558, 9540816132
E-mail: newindiapublishingagency@gmail.com
Website: www.nipaersources.com

Print ISBN: 978-93-58872-77-4

e-book ISBN: 978-93-58876-88-8

Composed and Designed by NIPA®.

GIS Mapping and Its Applications in Water Resources Management

K Ch V Naga Kumar
Scientist
Land and Water Management Research Group
KSCSTE- Centre for Water Resources Development and Management (CWRDM),
Kunnamangalam-673 571, Kerala

Naveena K
Scientist
Land and Water Management Research Group
KSCSTE- Centre for Water Resources Development and Management (CWRDM),
Kunnamangalam-673 571, Kerala

Surendran U
Principal Scientist
Division of Remote Sensing Application
ICAR-National Bureau of Soil Survey and Land Use Planning Nagpur
Vayusena, Nagar-440 033, Maharashtra

NIPA® GENX ELECTRONIC RESOURCES & SOLUTIONS P. LTD.
New Delhi-110 034

Decision Support systems for soil nutrient budgeting and its impact on crop yield, Adsorption of nutrient/Modeling nutrient behavior in soils and its chemistry, Carbon Sequestration in soils, Climate change and its impact on agriculture, Crop Simulation Modeling, etc. He has more than 100 national and international peer reviewed publications. He handled research projects in Natural Resources Maangement funded by various National and International Agencies.

Acknowledgement

Authors are thankful to the open source QGIS platform *(https://qgis.org/en/site/)* under which the demonstration of GIS exercises are peformed, and thanks also to earth explorer *(https://earthexplorer.usgs.gov/)* for giving access to the freely available satellite remote sensing data for visualizing and performing the analysis in the GIS environment.

Preface

Water is a valuable natural resource that is essential to all aspects of life on Mother "Earth" In fact, our planet is covered by 70% of water, but only 3% of the total water is in the form of freshwater, and again two-thirds of that is in the form of glaciers or is unavailable for our use. Presently, 1.1 billion people globally lack access to good drinking water resources, and another 2.7 billion find water scarce for at least one month of the year.

India has an annual average precipitation of 1,170 mm and about 80% of the total area experiences an annual rainfall of 750 mm or more, which is due to large spatial and temporal variability in the rainfall. With growing urban population and land use changes, water scarcity will be exacerbated, and also with climate change and bio-energy demands amplify the complex relationship between world development and water demand. The chapters cover the topic of:

Introduction to Remote Sensing and Its Applications

Quantitative Evaluation of Geomorphology Using Geo-Spatial Techniques: A Case Study of Chaliyar River Basin, India

Insights into the Groundwater Resource Mapping: Example from Kerala State, India

Assessing Spatial Changes in Blue and Green Water Resources: A Case Study of Valapattanam River Basin Using SWAT Model

GIS Tools for Mapping the Spatio-Temporal Trends in Rainfall Distribution: A Case Study of Kerala State, India

Geographic Information System (GIS): A Practical Guide Towards Data Visualization, Analysis and Map Preparation.

Editors

Contents

1

Introduction to Remote Sensing and Its Applications

Naga Kumar K.Ch. V[1], Surendran U.[2], Naveena K.[3] and Manoj P. Samuel[4]

[1,3-4]KSCSTE—Centre for Water Resources Development and Management (CWRDM) Kozhikode, India

[2]ICAR—National Bureau of Soil Survey and Land Use Planning (NBSS&LUP) Nagpur, India

Introduction

The term "Remote Sensing" was coined in 1950s by Ms. Evelyn Pruitt, a geographer with the U.S. Office of Naval Research, and the term is now commonly used to describe as the science and art of identifying, observing, and measuring an object on the earth surface, without having any direct contact. This process involves the detection and measurement of radiation of different wavelengths reflected or emitted from distant objects or materials through which they may be identified and categorized by class/type, substance, and spatial distribution *(National Aeronautics and Space Administration—NASA)*. Remote sensing technology has been widely applied in various applications such as water resources, agriculture, geological studies, forestry, land use, resource exploration, etc.

Elements of Remote Sensing

The remote sensing process involves an interaction between the incident radiation and the targets of interest, exemplified by the use of imaging systems in which the following seven elements are involved; however, remote sensing also involves sensing of emitted energy and using non-imaging sensors (Figure 1.1).

1. **Energy Source or Illumination (A):** The first requirement for remote sensing is an energy source that illuminates or provides electromagnetic (EM) energy to the target.

2. **Radiation and the Atmosphere (B):** As the energy travels from the source to the target and from the target to the sensor, energy interacts with the atmosphere.
3. **Interaction with the Target/Object (C):** Once the energy makes its way to the target through the atmosphere, it interacts with the target depending upon the properties of both the target and the radiation.

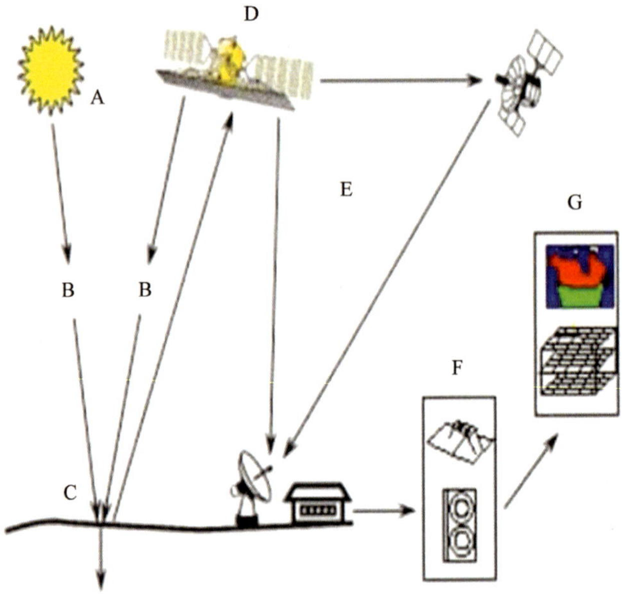

Fig. 1.1: Elements involved in remote sensing.

(Source: Canadian Centre for Remote Sensing; https://natural-resources.canada.ca/maps- tools-and-publications/satellite-imagery-and-air-photos/tutorial-fundamentals-remote-sensing/9309)

4. **Recording of Energy by the Sensor (D):** We need a sensor to collect and record the EMR scattered or emitted by the target.
5. **Transmission, Reception, and Processing (E):** The energy recorded by the sensor is transmitted (usually in electronic form) to a receiving station for processing into an image (hard copy and/or digital).
6. **Interpretation and Analysis (F):** The processed image is interpreted visually or digitally to extract information about the target.
7. **Application (G):** The final element of the remote sensing process is using the information about the target to understand it, reveal new knowledge or assist in solving a particular problem.

Electromagnetic Radiation

As discussed, the first requirement for remote sensing is an energy source to illuminate the target/object. This energy is in the form of EM radiation, which consists of an electrical field (E) and a magnetic field (M). Both these fields (Figure 1.2) travel at the speed of light (C) and there are two main characteristics of EM radiation, namely wavelength (λ) and frequency (Hz). The wavelength is the length of one wave cycle, which can be measured as the distance between successive wave crests. Frequency refers to the number of cycles of a wave passing a fixed point per unit of time.

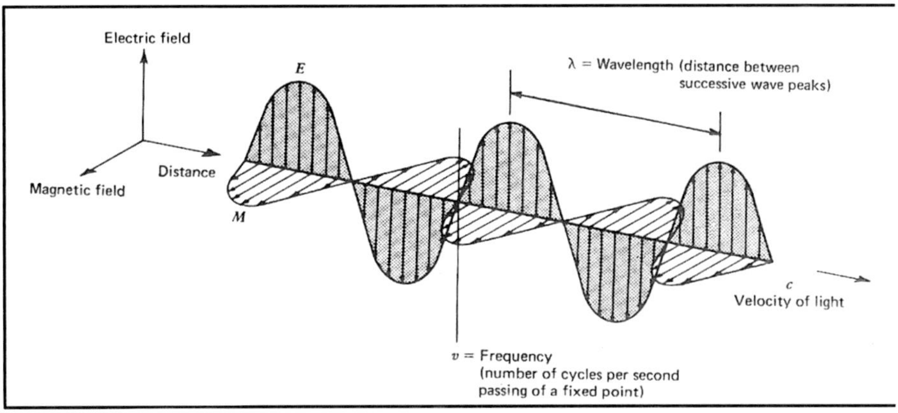

Fig. 1.2 : An electromagnetic wave. Components include a sinusoidal electric wave (E) and magnetic wave (M) at right angles, both perpendicular to the direction of propagation.

(*Source: Canadian Centre for Remote Sensing; https://natural-resources.canada.ca/maps-tools-&-publications/satellite-imagery-and-air-photos/tutorial-fundamentals-remote-sensing/9309)*

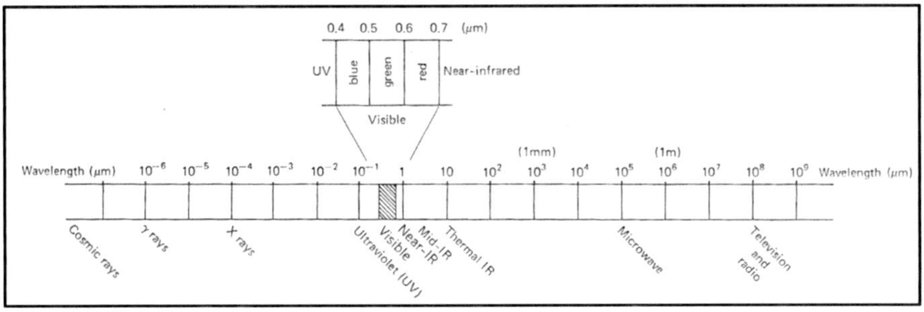

Fig. 1.3 : Electromagnetic spectrum

(*Source*: Canadian Centre for Remote Sensing)

Electromagnetic Spectrum

The EM spectrum consists of the ranges of all types of EM radiation, which is "energy" that travels and spreads out as it goes—the visible light that comes from a lamp in your house and the radio waves that come from a radio station are two types of EM radiation (Figure 1.3). The other types of EM radiation that make up the EM spectrum are microwaves, infrared light, ultraviolet light, X-rays, and gamma rays (NASA).

Spectral Reflectance

Remote sensing technology uses different regions of the EM spectrum namely, thermal, visible, infrared, and microwave. The objects on the land reflect a part of it to the atmosphere when interacting with these radiations which will be unique in nature. This reflectance can be used to identify and distinguish between the interaction materials. The spectral reflectance curve is the graph of spectral reflectance of objects as a function of wavelengths. Figure 1.4 shows the spectral reflectance curve of different types of land cover.

The reflectance of clear water is generally low. However, the reflectance is maximum at the blue end of the spectrum and decreases as wavelength increases. Turbid water has some sediment suspension that increases the reflectance in the red end of the spectrum. The reflectance of bare soil generally depends on its composition. Vegetation has a unique spectral signature that enables it to be distinguished readily from other types of land cover in an optical/near-infrared image. The reflectance is low in both the blue and red regions of the spectrum, due to absorption by chlorophyll for photosynthesis. It has a peak at the green region that is responsible for the green color of vegetation. In the near-infrared (NIR) region, the reflectance is much higher than that in the visible band due to the cellular structure in the leaves. Hence, vegetation

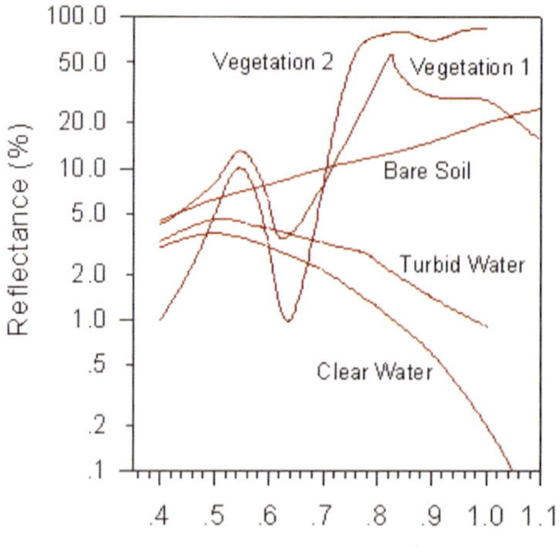

Fig. 1.4: Reflectance Spectrum of five types of land cover

(Source: https://crisp.nus.edu.sg/~research/tutorial/optical.htm)

can be identified by the high NIR but generally low visible reflectance *(https://crisp.nus.edu.sg/~research/tutorial/optical.htm)*.

Types of Remote Sensing

Based on the region used in the EM Radiation, the remote sensing systems can be categorized as follows:

a. **Optical remote sensing:** Optical remote sensing uses visible, infrared, and shortwave infrared portions of the EMR. The wavelength ranges from 0.4 to 3.75 µm. *(Example: Landsat MSS, Indian Remote Sensing (IRS) LISS III/IV, Sentinel MSS, IKONOS, MODIS, etc.)*
b. **Thermal Remote Sensing:** This type of remote sensing uses the thermal portion of the EM Spectrum which ranges from 3.75 to 12.5 µm. All objects emit thermal-infrared energy as heat and the thermal infrared sensors record this emitted energy. *(Example: ASTER, LANDSAT TIRS.)*
c. **Microwave Remote Sensing:** This type of remote sensing uses the microwave portion (1 cm to 1 m) of the EM spectrum. Because of the large wavelength, these sensors are capable of penetrating through clouds that give sensing all-weather capability *(Example: Sentinel SAR, NISAR, ALOS PALSAR)*.

Remote Sensing—Sensors and Platforms

Sensors: Instruments that measure EM radiation are called sensors. It can be classified as active sensors and passive sensors. Remote sensing systems which measure naturally available energy are called passive sensors. Accelerometers, hyperspectral radiometers, etc. are a few examples of passive sensors, and they work only under sunlight conditions. However, the active sensors use their own energy source for illumination and they record the reflected radiation from the target. They can work both day and night time (for example: Light Detection and Ranging (LiDAR) and Synthetic Aperture Radar (SAR) (https://www.nrcan.gc.ca).

Platforms: Platforms are used to carry the sensor. There are mainly three types of sensors based on their platform, i.e., Ground-based, Airborne, and Space-borne. The ground-based sensors are used for obtaining detailed information about the target compared to space-borne and airborne sensors, and they are placed on land-based vehicles, whereas airborne platforms include aero-planes/helicopters. The ground-based and airborne remote sensing are very costly and the area coverage is limited. The space-based sensors are mounted on the satellites. They can provide better area coverage and it is cost effective also. Apart from that, the other advantages of satellite remote sensing are its ability to synoptic view and easy integration with Geographic Information System (GIS).

Geostationary and Polar Orbiting Satellite

The path followed by a satellite is known as orbit. The satellites can be divided into two broad classes according to their orbits. They are:

Geostationary satellites: The satellites are placed on orbits at an altitude of about 36,000 km above the equator. These satellites will be able to see the same part of the Earth all the time since they move with the same velocity as Earth. All telecommunication and weather satellites are geostationary satellites and all have a very coarse resolution imagery. Geostationary Meteorological satellites (GMS) of Japan, Meteosat by Europe, and the INSAT series of India are some classical examples of Geostationary satellites.

Polar Orbiting Satellites: The satellites that follow the orbits inclining 90 degrees and altitude from 700 km to 800 km are called polar orbiting satellites. They can see the whole earth with higher resolution than that of geostationary satellites. IRS series satellites of India, LANDSAT series of USA, Sentinel series of Europe, etc. are Polar Orbiting Satellites.

Web Mapping Services

Web Mapping Services enables one to view and access different geospatial data in an online platform. Some of the online portals are:

- NRSC/ISRO BHUVAN *(https://bhuvan.nrsc.gov.in/home/index.php)*
- Water Resources Information System of India *(https://indiawris.gov.in/wris/#/)*
- India Biodiversity Portal *(https://indiabiodiversity.org/)*
- Survey of India (https://www.surveyofindia.gov.in/)
- Mahalanobis National Crop Forecast Centre (MNCRF) *(https://ncfc.gov.in/maps.html)*
- ICAR—National Bureau of Soil Survey and Land Use Planning *(http://www.bhoomigeoportal-nbsslup.in/)*

Remote Sensing-Based Atlas (in India)

Table 1.1 provides the data on remote sensing application projects.

Table 1.1: Remote sensing application projects

S.No.	Name of the Atlas	Source
1	River Basin Atlas	https://indiawris.gov.in/downloads/RiverBasinAtlas_Full.pdf
2	Wastelands Atlas	https://dolr.gov.in/en/documents/wasteland-atlas-of-india
3	Flood Hazard Zonation Atlas	https://ndma.gov.in/flood-hazard-atlases
4	Glacial Lake Atlas	https://www.nrsc.gov.in/sites/default/files/pdf/Announcements/GL_Atlas_Indus_12Nov2020_Final.pdf https://www.nrsc.gov.in/sites/default/files/pdf/Announcements/Glacial_Lake_Atlas_Ganga_Basin_NRSC.pdf https://www.nrsc.gov.in/sites/default/files/pdf/Announcements/Brahmaputra_GL_Atlas_nrsc.pdf https://www.nrsc.gov.in/sites/default/files/pdf/Announcements/IHR_GlacialLake_Atlas.pdf
5	Tea Garden Atlas	https://www.nrsc.gov.in/sites/default/files/pdf/Announcements/DIBRUGARH_TG_ATLAS.pdf https://www.nrsc.gov.in/sites/default/files/pdf/Announcements/JALPAIGURI_TG_ATLAS.pdf
6	Landslide Atlas	https://www.nrsc.gov.in/sites/default/files/pdf/Announcements/Landslides_Atlas_of_India_Updated_25Aug2023.pdf
7	Geology Atlas	https://www.nrsc.gov.in/sites/default/files/pdf/Announcements/Geology_Atlas_Book.pdf
8	Land Use Land Cover Atlas	https://www.nrsc.gov.in/sites/default/files/pdf/Announcements/LULC_Atlas_NRSC.pdf
9	Space-Based Observation of Indian Wetlands	https://vedas.sac.gov.in/static/pdf/wetland_book_final.pdf
10	National Wetland Inventory & Assessment (NWIA) Atlas	https://vedas.sac.gov.in/en/National_Wetland_Inventory_and_Assessment_(NWIA)_Atlas.html
11	Desertification Status Mapping (DSM) Atlas	https://vedas.sac.gov.in/en/Desertification_Status_Mapping_(DSM)_Atlas.html
12	Shoreline change Atlas	https://vedas.sac.gov.in/en/Shoreline_Change_Atlas.html
13	Snow cover Atlas of India	https://vedas.sac.gov.in/en/Snow_Cover_Atlas.html
14	Planetary Atlas	https://vedas.sac.gov.in/en/moon.html https://vedas.sac.gov.in/en/mars.html

S.No.	Name of the Atlas	Source
15	Polar Science Atlas	https://vedas.sac.gov.in/en/polarScience.html
16	Wave Atlas/ Samudra	https://vedas.sac.gov.in/static/pdf/wave_atlas.pdf

Remote Sensing Data Sources

- Bhoonidhi *(https://bhoonidhi.nrsc.gov.in/bhoonidhi/home.html)*
- Earth Explorer *(https://earthexplorer.usgs.gov/)*
- *Alaska Satellite Facility (https://asf.alaska.edu/)*
- *GLoVis (https://glovis.usgs.gov/)*
- NASA Earth Observations *(https://neo.gsfc.nasa.gov/)*
- ESA's Sentinel data *(https://sentinels.copernicus.eu/web/sentinel/home)*
- NASA Earth Data *(https://www.earthdata.nasa.gov/)*
- NOAA Class *(https://www.aev.class.noaa.gov/saa/products/welcome)*
- Meteorological & Oceanographic Satellite Data Archival Centre (MOSDAC) (https://www.mosdac.gov.in/)
- Met Office Hadley Centre Observations datasets *(https://www.metoffice.gov.uk/hadobs/)*
- NOAA Physical Sciences Laboratory (PSL) *(https://psl.noaa.gov/)*
- Nasa: Global Precipitation Measurements Mission (GPM) *(https://gpm.nasa.gov/)*
- G-LiHT *(http://gliht.gsfc.nasa.gov)*
- LVIS *(http://lvis.gsfc.nasa.gov)*

Remote Sensing and GIS Open-Source Software

- AdVance HYperspectral data Analysis Software (AVHYAS) *(https://sites.google.com/view/avhyas-sac-isro/home)*
- Microwave Data Analysis Software (MIDAS) *(https://vedas.sac.gov.in/en/download.html)*
- MIDAS RISAT 1A MRS Time Series Processor *(https://vedas.sac.gov.in/en/download.html)*
- The Sentinel Toolbox *(https://sentinels.copernicus.eu/web/sentinel/toolboxes/sentinel-1)*
- QGIS *(https://www.qgis.org/en/site/)*
- GRASS GIS *(https://grass.osgeo.org/)*
- PolSARPro *(https://earth.esa.int/eogateway/tools/polsarpro)*

References

https://natural-resources.canada.ca/maps-tools-and-publications/satellite-imagery-and-air-photos/tutorial-fundamentals-remote-sensing/9309
https://crisp.nus.edu.sg/~research/tutorial/optical.htm
https://www.nrcan.gc.ca
https://bhuvan.nrsc.gov.in/home/index.php
https://indiawris.gov.in/wris/#/
https://indiabiodiversity.org/
https://www.surveyofindia.gov.in/
https://ncfc.gov.in/maps.html
http://www.bhoomigeoportal-nbsslup.in/
https://indiawris.gov.in/downloads/RiverBasinAtlas_Full.pdf
https://dolr.gov.in/en/documents/wasteland-atlas-of-india
https://ndma.gov.in/flood-hazard-atlases
https://www.nrsc.gov.in/sites/default/files/pdf/Announcements/GL_Atlas_Indus_12Nov2020_Final.pdf
https://www.nrsc.gov.in/sites/default/files/pdf/Announcements/Glacial_Lake_Atlas_Ganga_Basin_NRSC.pdf
https://www.nrsc.gov.in/sites/default/files/pdf/Announcements/Brahmaputra_GL_Atlas_nrsc.pdf
https://www.nrsc.gov.in/sites/default/files/pdf/Announcements/IHR_GlacialLake_Atlas.pdf
https://www.nrsc.gov.in/sites/default/files/pdf/Announcements/DIBRUGARH_TG_ATLAS.pdf
https://www.nrsc.gov.in/sites/default/files/pdf/Announcements/JALPAIGURI_TG_ATLAS.pdf
https://www.nrsc.gov.in/sites/default/files/pdf/Announcements/Landslides_Atlas_of_India_Updated_25Aug2023.pdf
https://www.nrsc.gov.in/sites/default/files/pdf/Announcements/Geology_Atlas_Book.pdf
https://www.nrsc.gov.in/sites/default/files/pdf/Announcements/LULC_Atlas_NRSC.pdf
https://vedas.sac.gov.in/static/pdf/wetland_book_final.pdf
https://vedas.sac.gov.in/en/National_Wetland_Inventory_&_Assessment_(NWIA)_Atlas.html
https://vedas.sac.gov.in/en/Desertification_Status_Mapping_(DSM)_Atlas.html
https://vedas.sac.gov.in/en/Shoreline_Change_Atlas.html
https://vedas.sac.gov.in/en/Snow_Cover_Atlas.html
https://vedas.sac.gov.in/en/moon.html
https://vedas.sac.gov.in/en/mars.html
https://vedas.sac.gov.in/en/polarScience.html
https://vedas.sac.gov.in/static/pdf/wave_atlas.pdf
https://bhoonidhi.nrsc.gov.in/bhoonidhi/home.html
https://earthexplorer.usgs.gov/
https://asf.alaska.edu/
https://glovis.usgs.gov/
https://neo.gsfc.nasa.gov/
https://sentinels.copernicus.eu/web/sentinel/home
https://www.earthdata.nasa.gov/
https://www.aev.class.noaa.gov/saa/products/welcome
https://www.mosdac.gov.in/
https://www.metoffice.gov.uk/hadobs/

https://psl.noaa.gov/
https://gpm.nasa.gov/
http://gliht.gsfc.nasa.gov
http://lvis.gsfc.nasa.gov
https://sites.google.com/view/avhyas-sac-isro/home
https://vedas.sac.gov.in/en/download.html
https://sentinels.copernicus.eu/web/sentinel/toolboxes/sentinel-1
https://www.qgis.org/en/site/
https://grass.osgeo.org/
https://earth.esa.int/eogateway/tools/polsarpro

2

Quantitative Evaluation of Geomorphology Using Geo-Spatial Techniques: A Case Study of Chaliyar River Basin, India

***Prajith V.*[1] *and Rishma C.*[2]**

[1,2]Centre for Water Resources Development and Management (CWRDM) Kozhikode, Kerala, India

Abstract

Morphometric analysis plays a vital role in watershed management as the hydrological and geomorphological behaviors of a watershed are administered by its morphometric characteristics. A quantitative morphometric analysis will help to identify and assess the changes in a watershed associated with climatic change and anthropogenic activities. The remote sensing (RS) system and geographic information system (GIS) are considered to be the most viable methods for morphometric analysis as they require very little time and can be applied to a larger area. In this paper, the applicability of spatial techniques for morphometric analysis has been studied using the Chaliyar River basin as the case study. Morphometric parameters such as stream order, stream length, bifurcation ratio, drainage density, stream frequency, form factor, and circulatory ratio are calculated for the selected basin. The relationship between different morphometric parameters were also analyzed for better understanding. The study highlights the usefulness of spatial techniques in illustrating the hydrological response of a watershed through morphometric analysis.

Introduction

Watershed management has emerged as an important research area in recent years. Understanding the morphology of the catchment is one of the widely used methods for watershed management. Morphometry is the measurement and mathematical analysis of the configuration of the earth's surface, shape, and dimension of its landforms (Agarwal 1998; Obi Reddy et al., 2002). The study of watershed is very useful in drainage basin evaluation, silt erosion control, flood frequency analysis, watershed prioritization, natural resources management, and conservation (Waikar and Nilawar, 2014). The significance of morphometric analysis is highlighted in the basins that are either un-gauged or difficult to access. The morphometry is usually analyzed through measurement of linear, aerial, relief, gradient of channel network, and contributing ground slope of the basin (Magesh et al., 2012). The hydrological behavior of a basin can be easily determined through the areal parameters (Mahala, 2020.) The evolutionary history of any basin can be best understood through the implication of different relief morphometric measures of drainage basin (Sharma and Sarma, 2013).

GIS and RS techniques are proven convenient methods for morphometric characterization of sub-watersheds and prioritization of watersheds. As the satellite images provide a synoptic view of a large area, the RS and GIS methods become very useful in the analysis of drainage basin morphometry (Rai et al., 2017). Thus, the morphometric analysis using spatial techniques has emerged to be the fast and efficient method as compared to most of the conventional methods (Rao et al., 2010) such as manual extraction of drainage network and assigning the stream order from a published Survey of India topographic map. The present study aims at using remote sensing and GIS technology to compute various parameters of morphometric characteristics of the Chaliyar River basin.

Description About the Study Area

Chaliyar, the fourth longest river in Kerala originates from the Ilambaleri hills of Western Ghats, Wayanad, Kerala, and it discharges to the Lakshadweep Sea, the southern bank of which is known as Chaliyam and the northern bank as Beypore. The Chaliyar basin lies between Latitude (N) 11°08′ to 11°38′ and Longitude (E) 75°45′ to 76°35′. The River originates from the Ilambaleri hills in Gudalur Taluk of Nilgiri district in Tamil Nadu at an elevation of 2066 m above mean sea level. This inter-state river has a total drainage area of 2911 km^2 out of which 2535 km^2 lies in Kerala State and the remaining 388 km^2 area falls in Tamil Nadu. The river has a length of about 170 km. It flows through the Malappuram district for most of its length and then forms the boundary between Malappuram and Kozhikode districts for around 17 km. The six major tributaries Chaliyarpuzha, Punnapuzha, Kanjirapuzha, Karimpuzha, Cherupuzha, and Iruvazhinjipuzha constitute the Chaliyar river drainage system. Other important tributaries are Kurumanpuzha, Pandipuzha,

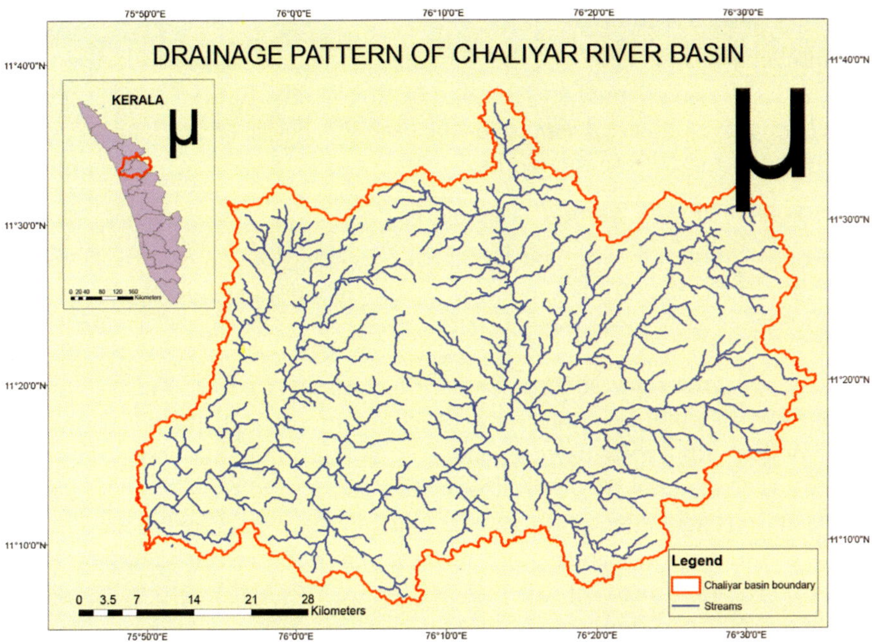

Fig. 2.1: Drainage map of Chaliyar river basin, India.

Maradipuzha, Kuthirapuzha, and Karakkodupuzha. Unlike many other rivers of Kerala, Chaliyar does not dry up during the summer season. Iruvazhinjipuzha is a major tributary of Chaliyar River and it joins Chaliyar at Koolimadu near Cheruvadi. The river originates from Vellarimala of Western Ghats. The river has a length of about 50 km. Its major tributary is Chalippuzha, which joins with it about 3 km north of Thiruvambady town. The villages of Anakkampoyil, Thiruvambadi, Mukkom, Kodiyaathoor, Chennamangaloor, and Cheruvadiare on the banks of this river. Figure 2.1 shows the location map of the Chaliyar River basin.

The Chaliyar basin drains through the soils having two types of runoff potential i.e., moderate low runoff potential and high runoff potential only. Most of the region in the basin is covered chiefly by clayey soils and these soils have a very low infiltration rate when it is thoroughly wetted. Hence, naturally, the swelling potential of the soils will be high and this property might have triggered the landslide/slip in the highland portions of the Chaliyar catchment especially in Kavalappara, Ambittampotty, etc. Also, the major portion of the catchment is covered with vegetation, especially in the upper reaches and the built-up part comes in the lower reaches of the river.

Materials and Methods

A comprehensive evaluation of the geometrical characteristics of a drainage basin and its stream channel system is to be done in the light of linear aspects of the channel system, areal aspects of the drainage basin, and relief aspects of drainage basin and channel networks.

Linear Aspects of the Channel System

Stream order, stream length, and length of overland flow constitute the most important variants that fall in the linear aspects of a drainage network.

Stream order

A very short and narrow stream channel that originates from the watershed divide is the basic building unit in the formation of a stream channel network. These smallest tributaries are designated as the first-order streams. Like this, two first-order streams that start from different locations of the watershed divide will join each other to form a new channel segment called second-order streams. Similarly, two second-order streams join to form a third-order stream, and so on. The ratio between the number of stream segments of given order N_u and the number of segments of the next higher order, N_{u+1} is termed as the bifurcation ratio R_b.

$$\boldsymbol{R_b} = \frac{\boldsymbol{N_u}}{\boldsymbol{N_{u+1}}}$$

For instance, a bifurcation ratio of 4 means that, the number of channel segments of a given order will be 4 times that of the next higher order. The bifurcation ratio of a network has a predominant role in the retention of flood extremes. An elongated basin having a high bifurcation ratio generates a low and extended peak flow compared to a normal basin having with low R_b which produces a hydrograph having a sharp peak. According to Horton, the number of stream segments of each order forms an inverse geometric relation with the order number and is hence called the law of stream numbers. i.e.,

$\boldsymbol{N_u} = \boldsymbol{R_b}^{k-u}$, where k is the order of the trunk segment.

Also, the logarithm of the number of stream segments follows a linear relationship with the stream order. The regression coefficient i.e., the slope of the linear regression line is a measure of the average value of the bifurcation ratio of the entire stream network. Horton introduced an expression to compute the total number of streams of all orders in a stream network as follows:

$$\sum_{i=1}^{k} N_{u=} \frac{R_b^{\,k} - 1}{R_b - 1}$$

Stream length

The characteristic size of components of a stream network and its contributing basin surfaces are closely related to the mean length of the stream channel segment of order *u*. The mean length of channel, $\overline{L}_u$ of order u is given by

$$\overline{L_u} = \frac{Total\ length\ of\ segments\ of\ order\ u}{Number\ of\ segments}$$

According to Horton's law of stream length, the mean lengths of stream segments of each of the successive orders of a basin tend to approximate a direct geometric sequence in which the first term is the average length of segments of the first order:

$$\overline{L_u} = \overline{L_1} R_L^{\,u-1}$$

where *represents* R_L represents the length ratio which is the ratio between the mean length of segments of order u and the mean length of segments of the next lower order. The validity of the law can be ascertained from the straight-line relationship between the logarithm of mean stream length as ordinate and stream order as abscissa. Horton also introduced the following expression to find out the total length of channel.

$$\sum_{i=1}^{N} L_u = \overline{L_1} R_b^{\,k-u} R_L^{\,u-1}$$

Furthermore, the total length of channels for all orders of a watershed of order *k* can be estimated as

$$\sum_{i=1}^{k} \sum_{i=1}^{N} L_u = \overline{L_1} R_b^{\,k-1} \frac{R_{Lb}^{\,k} - 1}{R_{Lb} - 1}$$

where, $R_{Lb} = \dfrac{R_L}{R_b}$.

Length of overland flow

Horton defined the length of overland flow *Lg* as the length of flow path, projected to the horizontal of non-channel flow from a point on the watershed divide to a point on the adjacent channel. As the number of starting points from the basin divide is infinite, an average length of overland flow is usually computed by taking measurements of flow paths emanating from points that are uniformly spaced around the entire basin perimeter. The maximum length of overland flow is the longest flow path traveled by a drop of water from the basin divide to the drainage outlet.

Areal Aspects of Drainage Basins

The area of basin corresponding to a given order of stream channel is the total area of land draining toward a particular location in that particular order stream channel. This area includes the drainage area of all lower-order streams that form the channel network and the area of land that directly contributes to the flow toward the drainage outlet (it is called inter basin area). Schumm expressed a relationship as the law of stream areas which states that the mean basin areas of streams of each order tend closely to approximate a direct geometric sequence in which the first term is the mean area of the first-order basin. This law is expressed as:

$$\overline{A_u} = \overline{A_1} R_a^{\,u-1}$$

where $\overline{A_u}$ is mean area of basins of order u, $\overline{A_1}$ is mean area of the first order basins and R_a is an area ratio analogous to the length ratio R_L. Further, Morisawa proved that there is a strong linear relationship exists between the logarithm of mean basin area of given order with the corresponding order of the stream channel. He also noted that the log-log plot between the mean stream length vs. area corresponding to a given order and the log-log plot between the cumulative total stream length vs. area holds a perfect linear relationship.

Basin shape factors

Basin shape or outline has a significant role in the rainfall-runoff characteristics of a basin. As already pointed out that the peak runoff and how immediately it happens after the rainfall event depends upon the elongated/circular nature of the basin. Horton put forward a dimensionless ratio to describe the drainage basin outline called *form factor*, R_f, which is the ratio between the basin area A_u and square of basin length L_b.

$$R_f = \frac{A_u}{L_b^2}$$

Further, Miller introduced an expression to understand the circular nature of a basin. This dimensionless *circularity ratio* R_c is defined as the ratio of basin area A_u to the area of a circle A_c having the same perimeter as the basin.

$$R_c = \frac{A_u}{A_c}$$

The elongation characteristics of a basin can be checked using the expression given by Schumm. He introduced the term elongation ratio as the ratio of the diameter of a circle of the same area as the basin to the maximum basin length.

Its value ranges between 0.6 to 1 and a value near to 1 means that the region of low relief.

Drainage Density

Drainage density, *D*, is the measure of the closeness of spacing of stream segments. Drainage density is defined (by Horton) as the ratio of the total length of streams of all orders within the basin to the basin area.

$$D = \frac{\sum_{i=1}^{k}\sum_{i=1}^{N} L_u}{A_u}$$

Constant of Channel Maintenance (C)

It is the area of drainage basin required to maintain one unit of channel length. It is the reciprocal of drainage density (Schumm 1956)

$$C = \frac{1}{D} = \frac{A_u}{\sum_{i=1}^{k}\sum_{i=1}^{N} L_u}$$

Strahler proved that a plot of Logarithm of basin area (ordinate) holds a linear relationship with the logarithm of the cumulative total stream length (abscissa) of particular order in the drainage basin. The constant of channel maintenance is the antilog of the y-intercept of this log-log plot.

Stream Frequency

Horton defined the stream frequency, F, as the number of stream segments per unit area of the drainage basin.

$$F = \frac{\sum_{i=1}^{k} N_u}{A_k}$$

where $\sum$ is the total number of stream segments of all orders within the given basin having an order of *k* and A_k is the area that basin.

Relief Aspects of Drainage Basins

Relief aspects deal with the three-dimensional parameters like basin relief, relief ratio, ruggedness number, river profile, gradient ratio, and hypsometric curve. Three relief aspects such as basin relief, relief ratio, and ruggedness number have been considered in this study.

Relief

Relief is the elevation difference between two reference points. Maximum relief (H) is simply the elevation difference between the highest and lowest points. It is usually the elevation difference between basin mouth and the highest point on the basin parameter, expressed in meters or feet.

$$H = hh - hl$$

where,

hh = Elevation of the highest point on the basin parameter

hl= Elevation of the basin mouth

Relief ratio

Relief ratio (Rh) is the ratio of maximum relief (H) to the horizontal distance on which it is measured. It is a dimensionless height-length ratio, with its higher value indicating a steep slope and high relief.

$$Rh = \frac{H}{LH}$$

Ruggedness number

Ruggedness Number (Rn) is the product of maximum relief (H) and the drainage density (D) (Schumm, 1956). To calculate Rn, both the parameters should be of the same unit. Rn indicates the geometric characteristics of a basin and its value will be extremely high when both the variables are large and the slope is steep. The low ruggedness value indicates less prone to soil erosion and vice versa.

$$Rn = H * D$$

Results and Conclusions

Figure 2.2 shows a schematic sketch of the distribution of stream segments in the Chaliyar watershed. It can be seen that the mainstream of the River is of fifth order. Naturally, the number of stream segments of any given order will be less in count when compared with the number of next lower order stream segments; however, it is more numerous than for the next higher order streams.

It is estimated that the total number of stream segments in the Chaliyar river basin with 1st, 2nd, 3rd, and 4th order is 433, 89, 21, and 5, respectively. Thus the bifurcation ratio between 1st and 2nd, 2nd and 3rd, 3rd and 4th, 4th and 5th order streams are 4.865, 4.238, 4.2, and 5, respectively. It means that each order of stream is bifurcated on average of more than 4 times in number to the next lower order. This study verified the law of stream order by fitting a regression equation between logarithm of number of stream segments with

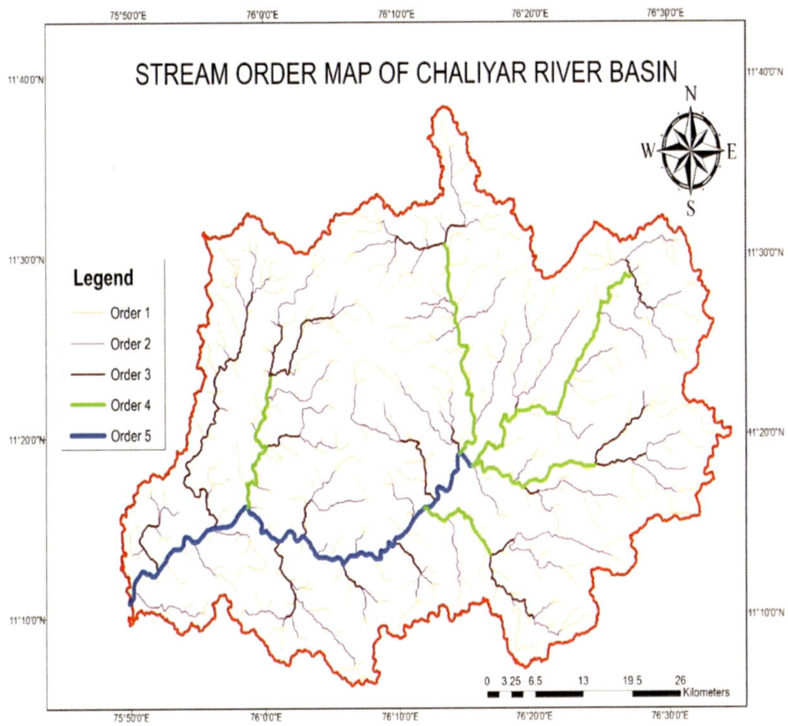

Fig. 2.2: Stream order map of the Chaliyar basin, India

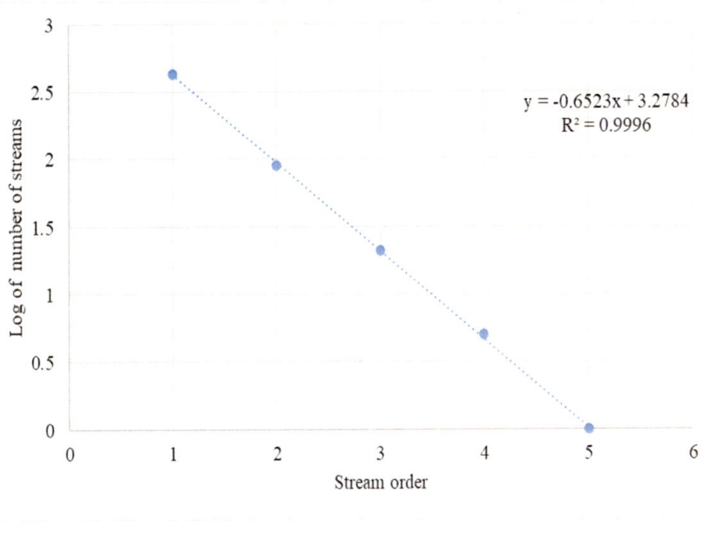

Fig. 2.3: Logarithm of number of streams vs Stream order

the stream order. The following plot (Figure 2.3) shows that these variables follow a linear relationship and hence the law of stream order is verified.

In Figure 2.3, the slope of the linear regression line represents the regression coefficient and hence the antilog of the regression coefficient gives the average value of bifurcation ratio of the entire stream network. The estimated value of the average bifurcation ratio of the Chaliyar watershed is 4.47. The bifurcated length of stream segments plays a significant role in the runoff characteristics of the river basin. It is obvious that the mean length of stream channel segments of a given order is greater than that of the next lower order but less than that of the next higher order. The study estimated that the total length of stream segments of all orders in the Chaliyar River basin is 1605.78 km. As it is an inter-state river, less than 10% of the stream length lies in the neighboring state of Tamil Nadu. The total length of 1st, 2nd, 3rd and 4th and 5th order streams is found as 803.827, 441.96, 122.913, and 67.77 km, respectively. However, the maximum length of overland flow is found as 123.155 km. Also, the following plot (Figure 2.4) shows the validity of the law of stream lengths from the straight line relationship between the logarithm of mean stream length as ordinate and stream order as abscissa.

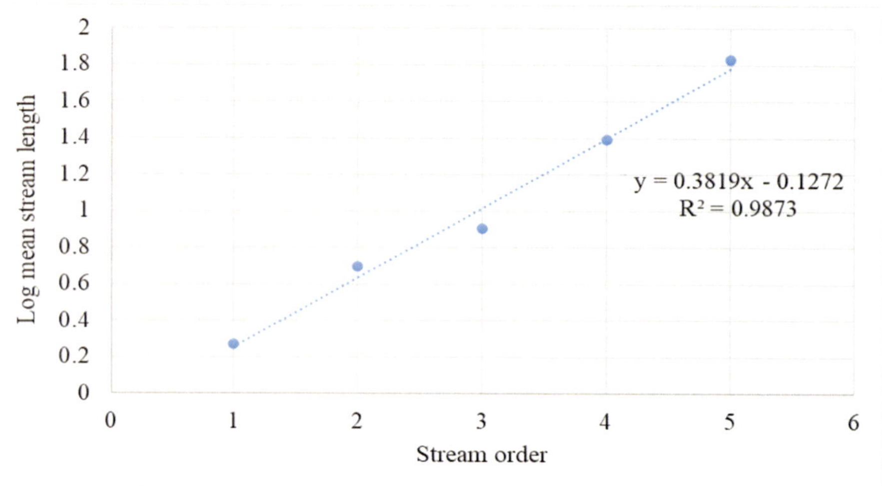

Fig. 2.4: Logarithm of mean stream length vs. Stream order.

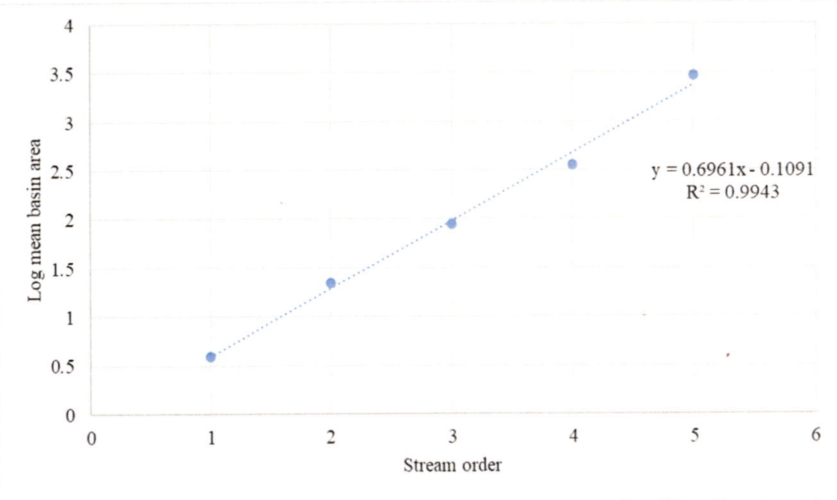

Fig. 2.5: Logarithm of mean basin area vs. stream order

This study tried to verify the law of stream areas by calculating the area of the basins of each order. It is found that the mean basin area of stream for each order tend closely to approximate a direct geometric sequence as shown in Fig. 2.5.

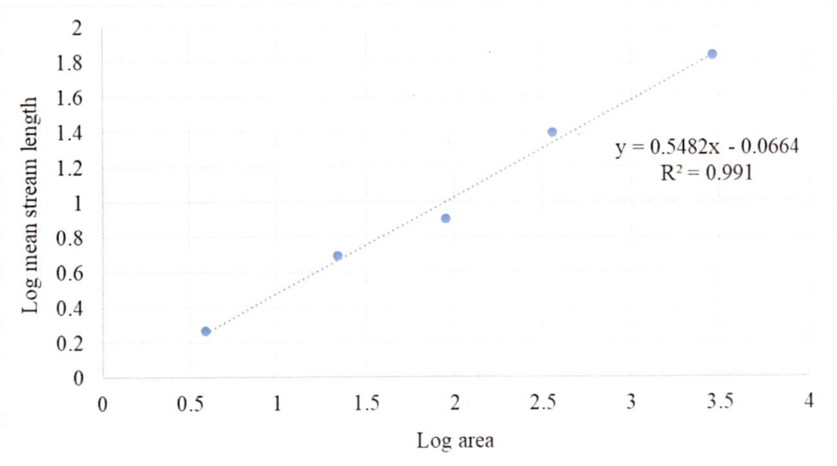

Fig. 2.6: Log-Log plot between mean stream length and basin area for different order

It is evident that there is a strong linear relationship exists between the logarithm of mean basin area of a given order with the corresponding order of the stream channel.

Also the log-log plot between the mean stream length vs. area corresponding to a given order holds a perfect linear relationship (Figure 2.6).

Coming to the area and relief aspects of the basin, the case study calculated the various variants such as Form factor, Circularity ratio, Elongation ratio, Drainage density, Constant of channel maintenance, Stream frequency, Relief, Relief ratio, and Ruggedness number. Table 2.1 lists various areal and relief morphometric parameters of the river basin.

The form factor ratio value for the study area is 0.407, which indicates the lower value of the factor and thus the basin will have low peak flow and longer duration. Comparing the circularity ratio and elongation ratio, it infers that the basin is more elongated and hence less probable for inundation. As the drainage density has vital role in the hydrological responses of a basin, a high value of this parameter is favored in regions having less or impermeable soil strata, sparse vegetation, and mountainous relief and a low value is favored in regions of highly permeable soil materials under thick vegetation cover. From the study, a low value of drainage density is obtained which indicates that the basin is an abstracting basin with sufficient vegetative cover. The low value of stream frequency also shows the thing as it is in positive relation with drainage density. The high value of the relief ratio indicates that the basin has hilly/ sloppy terrain and hence the basin is more susceptible to soil erosion when subjected to extreme rainfall events. The high value of ruggedness number reveals the same.

Table 2.1: Areal and relief parameters of Chaliyar River basin, India

Areal Parameters	
Form factor, r_f	0.407
Circularity ratio, r_c	0.263
Elongation ratio, r_e	0.72
Drainage density, D	0.551
Constant of channel maintenance, C	1.813
Stream frequency, f	0.188
Relief Parameters	
Relief, h (in m)	2620
Relief ratio (Rh)	32.737
Ruggedness number	1444.761

References

Agarwal CS. Study of drainage pattern through aerial data in Naugarh area of Varanasi district, U.P. Jour Indian Soc Remote Sensing 26,(1998): 169-175.

Magesh NS, Jitheshlal KV, Chandrasekar N, Jini KV. GIS based morphometric evaluation of Chimmini and Mupily watersheds, parts of Western Ghats, Thrissur District, Kerala. India Earth Sci Inform 5, no. 2, (2012):111-121.

Mahala A. The significance of morphometric analysis to understand the hydrological and morphological characteristics in two different morpho-climatic settings. Appl Water Sci 10, (2020): 33. https://doi.org/10.1007/s13201-019-1118-2

Obi Reddy GP, Maji AK, Gajbhiye KS. GIS for morphometric analysis of drainage basins. GIS India. 11, (2002): 9-14.

Rai PK, Mohan K, Mishra S, Ahmad A, Mishra VN. A GIS-based approach in drainage morphometric analysis of Kanhar River Basin, India. Appl Water Sci 7, (2017):217–232. DOI 10.1007/s13201-014-0238-y

Rao NK, Swarna LP, Kumar AP, Krishna HM. Morphometric analysis of Gostani River Basin in Andhra Pradesh State, Indian using spatial information technology. Int J Geomat Geosci 1, no. 2, (2010):179-187

Schumm SA. Evolution of drainage systems and slopes in badlands at Perth Amboy, New Jersey. Geol Soc Am Bull 67. (1956):597-646.

Sharma S, Sarma JN. Drainage analysis in a part of the Brahmaputra Valley in Sivasagar District, Assam, India, to detect the role of nontectonic activity. J Ind Soc Rem Sens 41(4):895-904.

Ven Te Chow. Handbook of Applied Hydrology. pp 4-40 to 4-69. (1988) New York: Mc Graw Hill.

Waikar M, Nilawar A. 2014. Morphometric Analysis of a Drainage Basin Using Geographical Information System: A Case study. 2.

3

Insights into the Groundwater Resource Mapping: Example from Kerala State, India

Raicy M.C.,[1] Arul T.[2], Aju C.D.[3] and Achu A.L.[4]

[1]Hydrology and Climatology Research Group, KSCSTE-Centre for Water Resources Development and Management, Kunnamangalam, Kozhikode, Kerala

[2]Department of Geology and Environmental Science, Christ College (Autonomous) Irinjalakuda, Thrissur, Kerala

[3]Centre for Climate Change Research, Indian Institute of Tropical Meteorology, Pune Maharashtra

[4]Department of Climate Variability and Aquatic Ecosystems, Kerala University of Fisheries and Ocean Studies (KUFOS), Kochi, Kerala

---

Abstract

Groundwater, a vital global resource, faces increasing stress due to rising demand, climate change, industrialization, and land use changes, which necessitates sustainable development strategies for its management. The present study aims to analyze the groundwater resource characteristics of Kerala, a southern state of India as a case study. The spatial variations in groundwater recharge, extraction, groundwater level, etc. in the state are discussed in this chapter. Seasonal average groundwater levels in the observation wells and rainfall were spatially represented in the GIS environment to analyze the spatial variations. Further, the river basin-wise variations in groundwater level were analyzed using box and whisker plots. The study shows that groundwater in the state is influenced by multiple factors such as rainfall patterns, topography, soil types, and human activities. Understanding the groundwater status is critical for decision-making as part of the sustainable development of groundwater resources in the state.

Keywords: *Groundwater, Extraction, India, Kerala, Recharge*

Introduction

Groundwater resources are under stress globally due to various reasons including increasing water demand, impending climate change, industrialization, land use/land cover changes, etc. Adequate sustainable development strategies have to be adopted for the management of groundwater resources. Irrespective of the developmental status of countries all over the world, groundwater serves as a crucial resource for meeting fundamental needs. However, increasing population, water demand, surface water contamination, and recent patterns of climate unpredictability have caused increasing reliance on groundwater as reported by several researchers (Adimalla and Taloor 2020; Aladejana et al. 2020; Cuthbert et al. 2019; Green 2016; Hamed et al. 2018; Jasrotia et al. 2019; Khalaj et al. 2019; Li 2016; McGill et al. 2019; Morsy et al. 2017; Taloor et al. 2020).

Even in India, water scarcity has been reported in many parts, though the country receives relatively higher rainfall. However, the rainfall pattern has changed over the last few years. The country has been experiencing significant spatiotemporal changes in rainfall patterns, leading to fewer rainy days and more extreme rainfall, which many studies have attributed to global climate change and anthropogenic influences. The rainfall variability has substantial impacts on water availability, agricultural productivity, and welfare of the 1.2 billion people in India because agriculture in India feeds about 17.2% of the global population and more than 56% of the agricultural land is rain-fed. Rapid and heavy patterns of rainfall result in the draining of most of the rainfall as surface runoff particularly on higher slopes, and reaches the surface water bodies, thereby reducing the infiltration and subsequent groundwater recharge. These variations in rainfall have ultimately affected the groundwater behavior and availability in many river basins.

As these challenges are expected to intensify in the future (Shaji and Pradeep 2019), it is necessary to implement sustainable development strategies that refer to the responsible management and use of groundwater resources to meet the current needs while ensuring their availability for future generations. The preliminary step in the sustainable development of the groundwater resources is to understand the current scenario of the groundwater system, estimate the groundwater resources, safe yield, and the possible methods of aquifer rejuvenation in the area. Such a study has been carried out in the state of Kerala, located in the southwestern part of India. Though the state receives an annual average rainfall of 3000 mm, the undulating topography of the state coupled with deforestation and sand mining in the rivers leads to an accelerated draining of water to the sea. This is evident from the fact that the groundwater recharging has suffered and groundwater levels have steeply declined in the state. As the main source of fresh water is groundwater, there is an increase in the extraction of aquifer water. Though the groundwater is replenished by the return of water from irrigation, most of the cultivation fields have been consumed for the construction of large buildings which caused a reduction in the aquifer recharge percentage (George 2016). The rainfall pattern in the state has changed abruptly since the 1990s (Nair et al. 2021). The heavy and short-period downpours cause most surface runoff to drain faster and reduce the groundwater recharge. In addition, the typical east-to-west sloping physiography of the state influences the drainage. Of the available water resources, because of sloping only 58.54% is used and the rest is drained into the sea as the state slopes toward the west. In this context, the present study aims to review and analyze the status of groundwater resources in Kerala, which would help in identifying the areas where groundwater management is necessary.

Study Area

Kerala state covers an area of 38,863 sq. km and is a narrow stretch of land located in the southwestern part of the Indian Peninsular Shield (Figure 3.1). This linear strip of land is bounded by the Western Ghats on the east and the Arabian Sea on the west. Physiographically, the state is divided into lowlands *(less than 7.6 m amsl)*, midlands *(7.6–76 m amsl)*, and highlands *(more than 7.6 m amsl)*. The state is also divided into four broad geomorphologic units, namely (i) the coastal strip fringing the Lakshadweep Sea, (ii) the landforms marked by laterite capping between altitudes of 30 and 200 m, (iii) the foothills of Western Ghats ranging in altitude from 200 to 600 m and (iv) the steeply rising Western Ghats hill ranges with altitudes reaching up to 2500 m. Of the total area of the state, 35,955 sq. km area is constituted by crystalline and the rest by soft sediments. The former comprises charnockites, gneiss, granite, meta-sediments, gabbro, and dolerite to mention the major ones. The sedimentary formations occur mostly in the coastal areas. Mineral deposits of clay, bauxite, rare earth sands, glass sand, iron ore, limestone, gold, graphite, chrysoberyl, etc. are known to occur in the state. The state accounts for 1.18% of the land area, 3.44% of the total population, and 3.58% of the water resources in the country.

Climate

The state undergoes an equatorial tropical climate and displays a variety of climatic conditions owing to its diverse geographical features. The eastern highlands, encompassing the mountainous areas, primarily feature a cool climate, while the coastal plains of the western lowlands generally remain hot and humid. The average annual rainfall in Kerala is quite high at 3000 mm per year with the monsoon period extending for four months. There are four seasons in Kerala, which include winter, summer, south-west monsoon, and north-east monsoon, among which south-west monsoon and north-east monsoon give a higher amount of rainfall. The southwest monsoon typically concludes by late September with an average rainfall ranging from 2250 to 2500 mm. While the northeast monsoon often referred to as the retreating monsoon, commences in October and extends through November. This period remains warm and muggy, with an average rainfall ranging from 450 to 500 mm, and is characterized by thunderstorms, lightning, and substantial precipitation.

Methodology

As part of the study, secondary data were collected from various agencies and the field. A geodatabase was prepared by digitizing the topographical contours

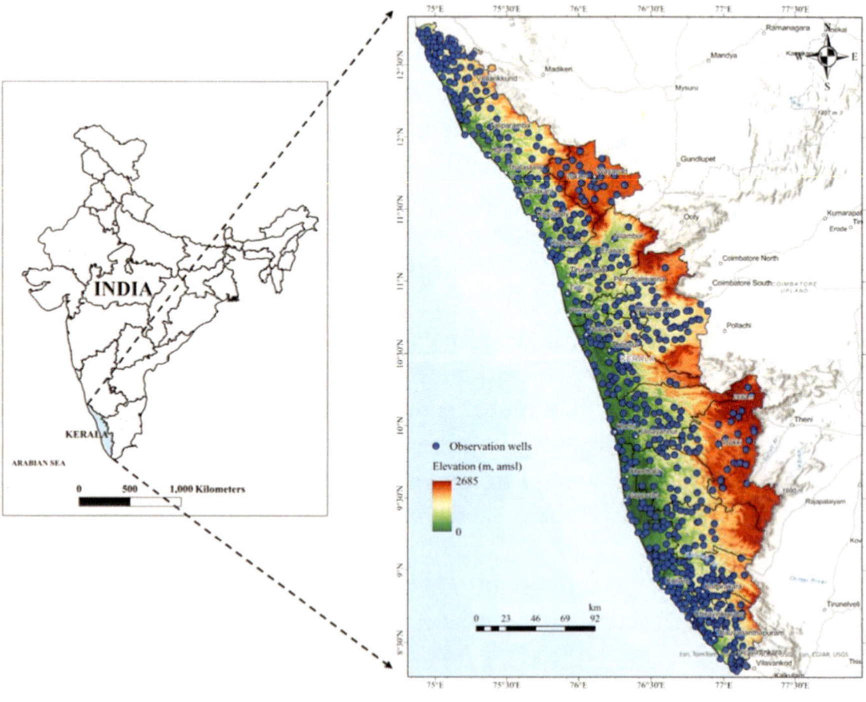

Fig. 3.1: Location of the study area with DEM and well locations (CGWB, GoI)

of toposheets at 1:50,000 scale through on-screen digitization in GIS, and the contour map was interpolated to prepare the Digital Elevation Model (DEM). The drainage map was derived for the study area from the toposheets; the drainage network was analyzed using Horton's scheme and Strahler's stream orderings (Strahler 1957); and the study area was extracted using the ArcGIS clip tool. The surface water and groundwater resource information of the state were accessed from various literature (CGWB 2022; SGWD 2020; Verma 2017). The groundwater level data for the year 2020 was accessed from the Central Groundwater Board (CGWB) and the seasonal groundwater levels were spatially interpolated using GIS software.

Surface Water Resources

The state is drained by 44 rivers (Figure 3.2), among which Kabini, Bhavani, and Pambar originate in the Western Ghats and are east-flowing, while others are west-flowing joining the Arabian Sea. Four rivers are medium categorized while others are minor category rivers as per the national norm (Rao 1979). Southwest monsoon and the Northeast monsoon play a significant role in shaping the hydrological characteristics of the river basins, with heavy rainfall during the monsoon seasons replenishing the rivers and sustaining the freshwater ecosystems. The river basins and drainage distribution in the state are shown in Figure 3.2. Periyar River basin is the largest river basin covering a significant part of the central and southern regions of the state, and the Manjeswaram River basin is the smallest basin. State Irrigation Board, Kerala (SIB 2023) reported that nearly 55,734 water bodies are present in the State among which 91.5% are ponds, and the remaining include tanks, reservoirs, water conservation schemes, streams, etc. The highest number of ponds and reservoirs are present in Palakkad and Pathanamthitta districts, respectively (SIB 2023).

Groundwater Resources

The principal aquifers in Kerala are crystalline, tertiary, lateritic, and alluvial aquifers and their general characteristics are shown in Table 3.1. In regions characterized by hard rock terrain, comprising weathered crystallines and laterites, groundwater distribution is complex. Within these formations, groundwater is typically found in phreatic conditions within the weathered residuum and shallow fractures, and semi-confined to confined conditions in deeper fracture zones. On the other hand, alluvial terrain typically hosts groundwater in shallow phreatic systems. The Tertiary sedimentary formations underlying certain areas harbor potential aquifers, often confined to semi-confined conditions. These aquifers, particularly prevalent in granular zones, are significant sources of groundwater, commonly accessed through dug wells. Notably, Charnockite rocks within the region are identified as superior aquifers compared to the Khondalite group, with more than 40% of wells in these formations yield substantial amounts of water.

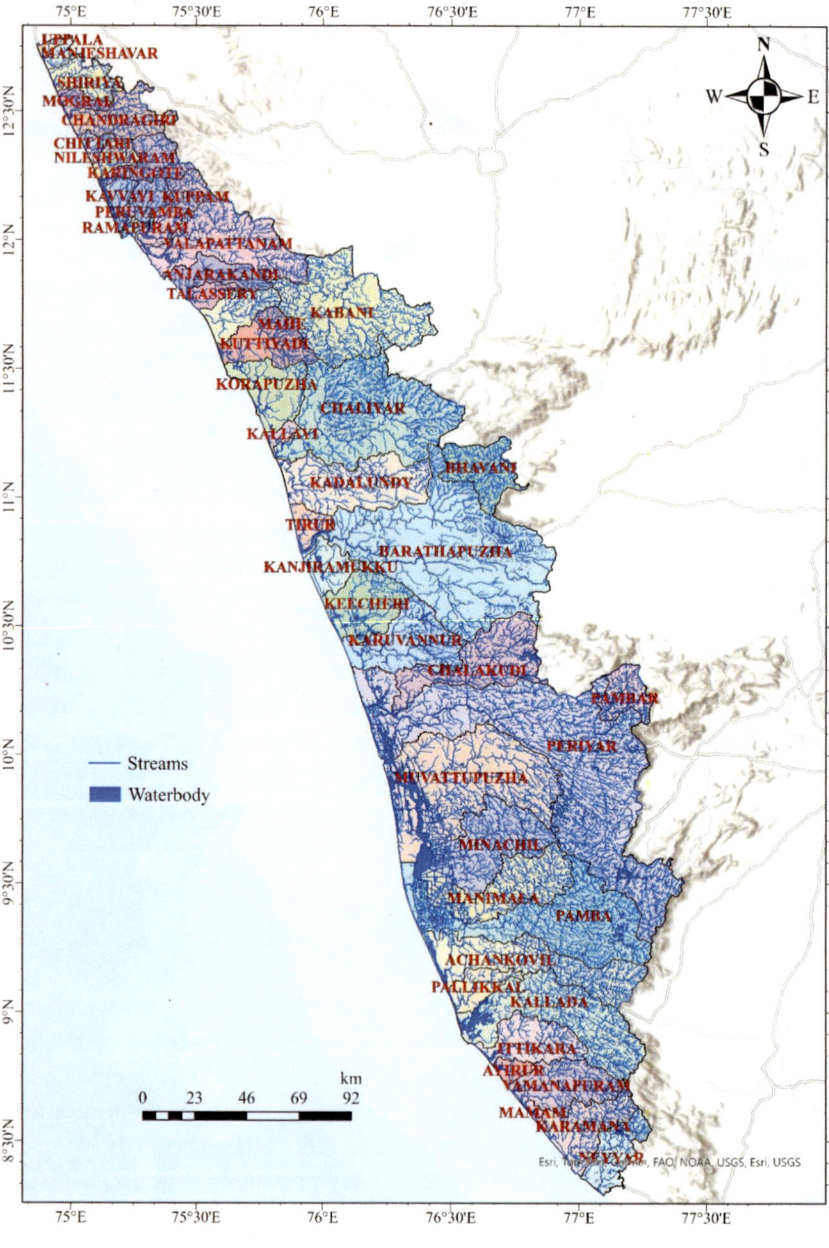

Fig. 3.2: The river basins of Kerala with streams, rivers, and major drainage distribution

Along the Kerala coast, the Tertiary formations are divided into Alleppey, Vaikom, Quilon, and Warkali beds (Kunhambu et al. 2018). While the Vaikom aquifers are generally productive, providing essential water resources, groundwater in these formations can be highly mineralized, especially in specific regions like the Kuttanad area and coastal zones west of Vembanad Lake (CGWB Report 2013). Laterites, widely distributed across the state, serve as potential aquifers, particularly along topographic lows and valleys. However, due to their porous nature, groundwater dynamics in laterites are heavily influenced by topography. Additionally, alluvial deposits along coastal plains are significant groundwater reservoirs, extensively utilized through dug wells and filter point wells. The depths to water level across these formations vary considerably, influenced by factors such as topography, rainfall patterns, and the thickness of overburden. Despite the abundance of groundwater resources, challenges such as mineralization in specific regions and fluctuations in water levels due to varying geological and hydrological conditions underline the importance of comprehensive management and monitoring of groundwater resources in the region.

District-wise recharge

The analysis of groundwater recharge dynamics across the districts of Kerala reveals notable variations in recharge volumes, sources, and seasonal patterns as shown in **Figure 3.3**. Trivandrum stands out with the highest recharge from rainfall during the monsoon season, reaching 201.6 MCM, while Wayanad records the lowest at 181.6 MCM, showcasing the variability in rainfall distribution and its impacts on groundwater replenishment. Conversely, Palakkad demonstrates the highest recharge from other sources during the monsoon season, at 41.97 MCM, underscoring the importance of surface water bodies in augmenting groundwater reserves, particularly in regions susceptible to intense rainfall events and runoff. During non-monsoon seasons, Thrissur emerges as a district with substantial recharge from both rainfall (479.94 MCM) and other sources (144.65 MCM), indicating its resilience to sustain groundwater replenishment beyond the traditional rainy months. In contrast, Wayanad displays the lowest recharge from both rainfall and other sources during the non-monsoon season, highlighting challenges related to seasonal water scarcity and the need for targeted interventions to enhance recharge mechanisms in water-stressed areas.

Examining the total annual groundwater recharge provides a comprehensive perspective on cumulative replenishment. Thrissur leads with the highest total annual groundwater recharge at 649.63 MCM, reflecting robust recharge mechanisms and favorable hydrological conditions. In contrast, Idukki reports the lowest total annual groundwater recharge, amounting to 210.07 MCM, indicating challenges related to groundwater availability and sustainability. These findings underscore spatial heterogeneity in recharge patterns and emphasize the importance of localized water management approaches tailored to address regional variations. Moreover, Kannur demonstrates a significant total annual groundwater recharge at 456.69 MCM. Furthermore, Ernakulam exhibits notable recharge from both rainfall (363.73 MCM) and other sources (70.21 MCM) during the monsoon season, contributing to its total annual groundwater recharge of 509.34 MCM.

Table 3.1: The types and characteristics of various aquifers in Kerala. (CGWB Report 2022; SGWD Report 2020)

Aquifer Type	Precambrian Crystalline (Hard Rocks) Aquifer		Tertiary (Alluvial zones) Aquifer		Laterite (Hard Rocks) Aquifer	Alluvial Aquifer
Formation	Shallow weathered zone	Deep fracture zone	Shallow aquifer	Deep granular zone	Shallow zone	Shallow weathered and deep zones
Depth to water level	2–16 mamsl	30–175 mbgl	2.0–27 mbgl	-	2–25 mbgl	<1–6 mbgl
Yield	2–10 m^3 per day	< 1–35 lps	500 lpd–10 m^3/day	Up to 49 lps	0.5–30 m^3/day	5–35 m^3/day
Groundwater occurrence	Phreatic	Semiconfined to Confined	Phreatic	Confined to semiconfined	Phreatic	Phreatic and semi-confined
Occurrence	Midlands, Hill ranges		Vaikom and Warkali beds	Alleppey bed	Midlands, parts of lowlands	Midlands, coastal plains, highlands
Potential aquifers	Charnockites		Vaikom beds, Warkali beds		Topographic lows	Coastal plains

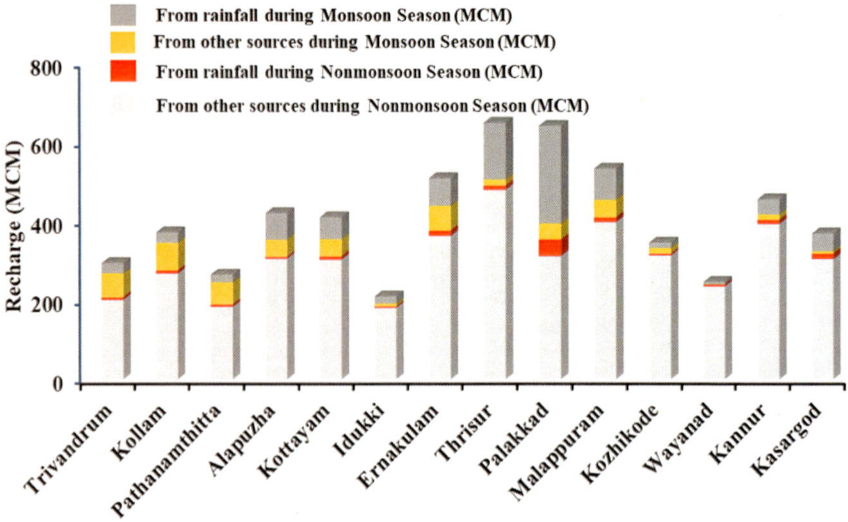

Fig. 3.3: The district-wise sources of recharge in Kerala for the year 2022. (CGWB report 2022)

District-wise groundwater extraction

Analysis of groundwater dynamics across various districts in the state offers valuable insights into the usage patterns, sustainability, and potential disparities. The district-wise extraction of groundwater for various purposes is shown in Figure 3.4. Analysis of the data reveals significant variations in groundwater recharge, extraction, and remaining balance among districts. For instance, while Palakkad is characterized by the highest annual recharge of 589.82 MCM, Idukki exhibits the lowest (189.06 MCM). Such disparities extend to extraction rates, with Kasargod recording the highest annual extraction for irrigation at 169.21 MCM, while Wayanad reports the lowest at 13.56 MCM. Interestingly, Pathanamthitta demonstrates no industrial extraction, contrasting with the maximum industrial usage of 6.06 MCM at Palakkad. Similarly, domestic usage varies substantially, with Malappuram consuming the most groundwater at 241.43 MCM, and the least at Kottayam at 0.03 MCM. Considering total extraction, Palakkad again leads with 339.35 MCM, while Wayanad trails with 55.46 MCM. Out of the total annual extractable groundwater recharge, Kottayam maintains the highest at 235.3 MCM. The findings show that there is a need to promote sustainable usage for safeguarding groundwater resources by implementing site-specific interventions.

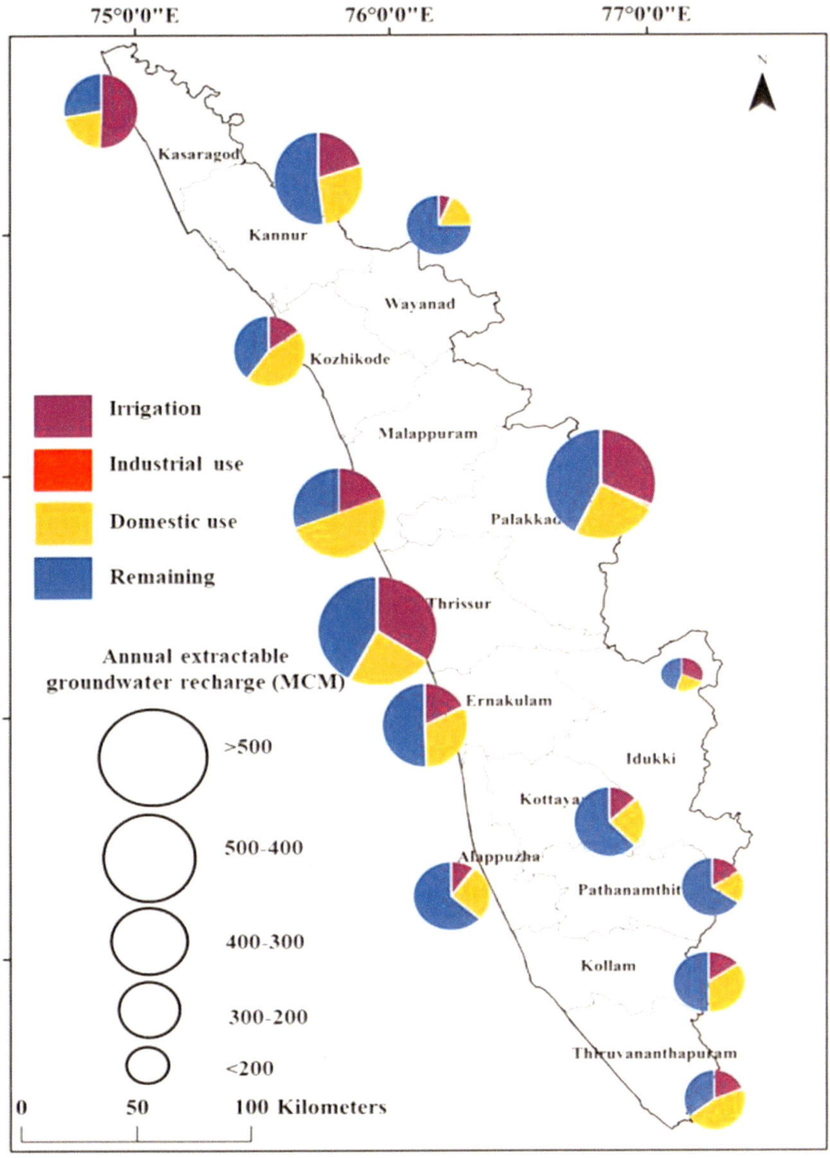

Fig. 3.4: Annual extractable groundwater recharge for all the districts for the year 2022. (Modified from CGWB 2022)

Spatial Variations in Groundwater Level

The precipitation rate and groundwater level in the state vary both seasonally and spatially and the spatiotemporal variations in the precipitation and the groundwater level are shown in Figure 3.5. In 2020, Kerala experienced diverse rainfall patterns throughout its four seasons. From January to March, the southern region experienced the highest rainfall, correlating with groundwater levels ranging from a minimum of 1 to a maximum of 14 mbgl. Conversely, the northern region, receiving less rainfall, showed lower groundwater levels, ranging from 5 to 14 mbgl, with most areas between 5 and 10 mbgl. During April to June, the southern and northern regions received comparable rainfall, while the central region experienced less. The groundwater levels in the northern region ranged from 5 to 14 mbgl, contrasting with the January to March season when most areas fell between 5 and 10 mbgl. However, during April to June, the majority of the northern region recorded levels between 11 and 14 mbgl, with few areas at 5 to 10 mbgl. In the south, levels ranged from 1 to 10 mbgl, with hilly areas at 11 to 14 mbgl due to significant runoff. Despite lower rainfall in the central part, most areas had levels between 11 and 14 mbgl. During July to September, the northern region received the most rainfall, while the south and central regions witnessed less or equal precipitation. Overall, the state's aquifers were saturated with water levels ranging from 1 to 10 mbgl, indicating uniform and abundant rainfall across districts, with the highest of the year.

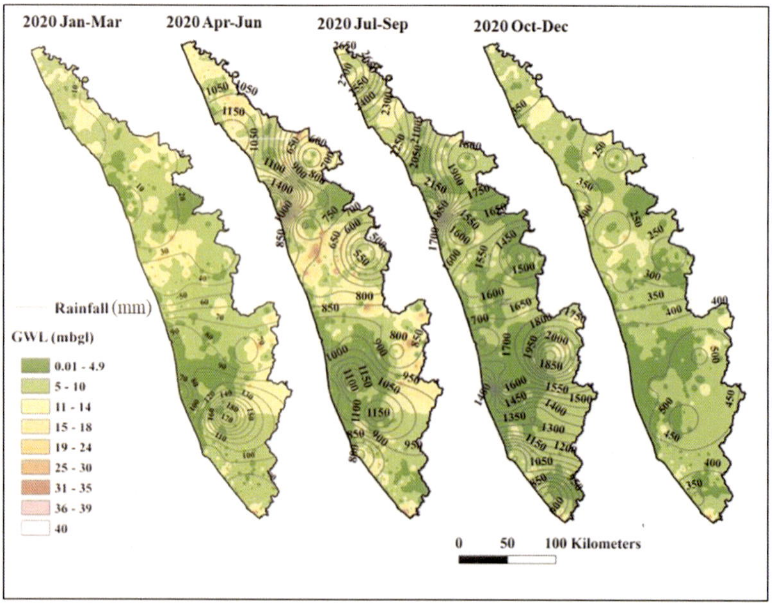

Fig. 3.5: Spatial variations in seasonal average groundwater level with rainfall contours for 2020

The runoff and evaporation rates were low, allowing for adequate absorption by underground storage systems. During October to December, the state experienced shallow and evenly distributed rainfall, resulting in groundwater levels between 1 and 10 mbgl. Despite lower precipitation compared to the previous season, water penetrated the aquifers instead of running off to the ocean due to lower temperatures and shallow downpours.

River Basin-Wise Groundwater Level Variation

Based on the groundwater level data for 2020, the river basin-wise groundwater levels with their spatial variations are shown by box and whisker plots (Figure 3.6) and the inference is as follows. The water levels exhibit

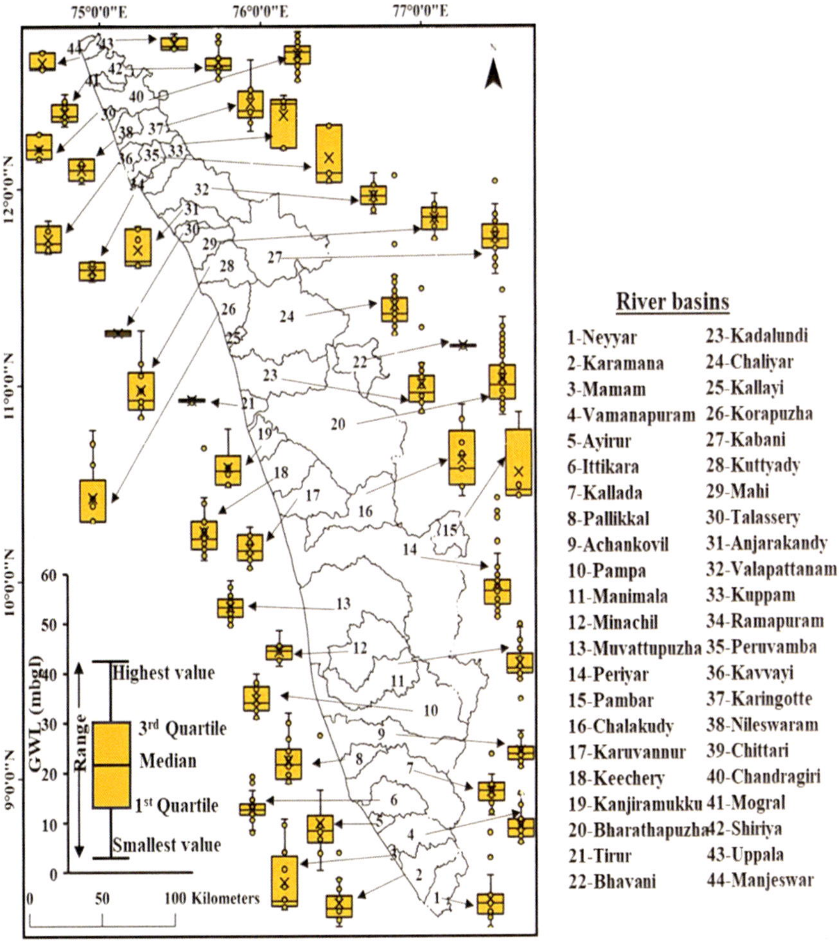

Fig. 3.6: The river basin-wise statistics of groundwater levels in 2020

substantial variation, ranging from as low as 0.09 m in the Mamom River Basin to 33.05 m in the Ayirur River Basin. Such diversity underscores the distinct hydrological characteristics of each basin. Higher groundwater level variability is observed in the Mamam, Pampar, Chalakudy, and Peruvamba River basins while lower groundwater level variability is observed in the Tirur, Thalassery, and Bhavani River basins. Some basins, like the Karamana and Vamanapuram, maintain relatively stable water levels (between 5.81 and 8.36 m). In contrast, the Ayirur basin experiences significant fluctuations, possibly due to seasonal factors or anthropogenic influences. The Neyyar basin records the highest water level of 17.3 m, while the Ithikkara basin has the lowest recorded water level of 0.1 m. The Neyyar River basin exhibits an average water level of 9.08, with a maximum of 27.74 and a minimum of zero, indicating considerable variability with a standard deviation of 6.56. In contrast, the Karamana River basin shows an average water level of 7.77, with a maximum of 19.21 and a minimum of zero, suggesting slightly lower variability with a standard deviation of 5.92. Groundwater needs to be managed artificially in areas with higher temporal groundwater level fluctuations in many areas particularly in parts of Palakkad, Kozhikode, Malappuram, Wayanad, Kasargod, and Thiruvananthapuram districts, where water scarcity is observed during summer. The diverse topography, soil types, climate conditions, and human activities such as irrigation, urbanization, groundwater extraction, etc. in each basin contribute to these disparities. Based on the disparities in groundwater availability in various river basins, there is a need for sustainable development of groundwater resources particularly in water stress zones.

Conclusions

This chapter examines the groundwater characteristics of Kerala and the major inferences are shown below.

- Groundwater in Kerala is influenced by factors such as rainfall patterns, topography, soil types, and human activities.
- Seasonal variations in rainfall significantly influence groundwater levels across Kerala, with the southern region experiencing higher groundwater levels correlating with greater rainfall, while the northern region displays lower groundwater levels.
- The groundwater level data for 2020 across various river basins in Kerala highlights significant spatial variations, with diverse hydrological characteristics influenced by factors such as topography, soil types, climate conditions, and human activities, underscoring the necessity for tailored management strategies to address fluctuations and water scarcity issues in specific districts.

The study emphasizes the need for comprehensive management strategies to address spatial and temporal disparities in groundwater availability in Kerala.

References

Adimalla N, Taloor AK. Hydrogeochemical investigation of groundwater quality in the hard rock terrain of South India using geographic information system (GIS) and groundwater quality index (GWQI) techniques. Groundw Sustain Dev, 10, (2020): 100288. (https://doi.org/10.1016/j.gsd.2019.100288)

Aladejana JA, Kalin RM, Sentenac P, Hassan I. Assessing the impact of climate change on groundwater quality of the shallow coastal aquifer of eastern Dahomey Basin Southwestern Nigeria. Water, 12, no. 1, (2020):224. (https://doi.org/10.3390/w12010224)

C.D. Aju, Achu A L, Mohammed Maharoof P, M.C. Raicy, Rajesh Reghunath, Girish Gopinath, (2024) Emerging nitrate contamination in groundwater: Changing phase in a fast-growing state of India, Chemosphere,357.

Cuthbert MO, Gleeson T, Moosdorf N, Befus KM, Schneider A, Hartmann J, Lehner B (2019) Global patterns and dynamics of climate–groundwater interactions. Nat Clim Chang, 9, no. 2, (2019): 137-41. (https://doi.org/10.1038/s41558-018-0386-4)

Central Ground Water Board, Ministry of Water Resources "Groundwater information booklet of Alappuzha district, Kerala State", 2013. (https://www.cgwb.gov.in/old_website/District_Profile/Kerala/Alappuzha%20.pdf)

Central Ground Water Board, Ministry of Water Resources Report, 2022. "Groundwater Resources of Kerala".

George R M. Bore Wells Vs. Open Wells: Water Crisis and Sustainable Alternatives in Kerala. Journal of Management & Public Policy, 7, no. 2, (2016): 19-28.

Ground Water Department Government of Kerala & Central Ground Water Board, Government of India "Dynamic Ground Water Resources of Kerala", 2020. (https://groundwater.kerala.gov.in/wp-content/uploads/2022/07/Groundwater-resource-Estimation-Committee-Report-2020.pdf)

Green TR. 2016. Linking climate change and groundwater. In: Jakeman AJ, Barreteau O, Hunt RJ, Rinaudo J-D, Ross A (eds). Integrated groundwater management. pp 97–141. Cham: Springer. (https://doi.org/10.1007/978-3-319-23576-9_5)

Hamed Y, Hadji R, Redhaounia B, Zighmi K, Bâali F, El Gayar A. Climate impact on surface and groundwater in North Africa: a global synthesis of findings and recommendations. Euro-Mediterr J Environ Integr, 3, no. 1, (2018): 25. (https://doi.org/10.1007/s41207-018-0067-8)

Jasrotia AS, Taloor AK, Andotra U, Kumar R. Monitoring and assessment of groundwater quality and its suitability for domestic and agricultural use in the Cenozoic rocks of Jammu Himalaya, India: a geospatial technology based approach. Groundw Sustain Dev, 8, (2019): 554-66. (https://doi.org/10.1016/j.gsd.2019.02.003)

Khalaj M, Kholghi M, Saghafian B, Bazrafshan J. Impact of climate variation and human activities on groundwater quality in northwest of Iran. J Water Supply Res Technol AQUA, 68, no. 2, (2019): 121-35 (https://doi.org/10.2166/aqua.2019.064)

Kunhambu V, Sureshbabu D.S and Vinayachandran N. Hydrogeo-Stratigraphic Model of Coastal Sediments in Kuttanad Area of Kerala, India. Nature Environment and Pollution Technology, 18, no.1, (2018): 299-305.

Li R. 2016. Assessing groundwater pollution risk in response to climate change and variability. In: Fares Al (ed) Emerging issues in groundwater resources. Pp 31-50. Cham: Springer. (https://doi.org/10.1007/978-3-319-32008-3_2)

McGill BM, Altchenko Y, Hamilton SK, Kenabatho PK, Sylvester SR, Villholth KG. Complex interactions between climate change, sanitation, and groundwater quality: a case study from Ramotswa Botswana. Hydrogeol J, 27, no. 3, (2019): 997–1015. (https://doi.org/10.1007/s10040-018-1901-4)

Morsy KM, Alenezi A, AlRukaibi DS. Groundwater and dependent ecosystems: revealing the impacts of climate change. Int J Appl Eng Res, 12, no. 13, (2017): 3919-26 (https://www.researchgate.net/profile/Karim-Morsy/publication/318929478_Groundwater_and_Dependent_Ecosystems_Revealing_the_Impacts_of_Climate_Change/links/59861e2245851560584a379b/Groundwater-and-Dependent-Ecosystems-Revealing-the-Impacts-of-Climate-Change.pdf)

Nair A, Nair V, Nair S. The role of North West Equatorial Indian Ocean in modulating the variability of southwest monsoon rainfall over Kerala in a climate change perspective. Theoretical and Applied Climatology, (2021): 146. 10.1007/s00704-021-03716-8.

Strahler A N. "Quantitative Analysis of Water Shed Geomorphology", Trans American Geophysical Union, 38, (1957): 913-20.

State Ground Water Department Government of Kerala. 2020. "Groundwater Resource Committee Report".

Swain, S., Taloor, A.K., Dhal, L. *et al.* Impact of climate change on groundwater hydrology: A comprehensive review and current status of the Indian hydrogeology. Appl Water Sci, 12, no. 6, (2022): 120. (https://doi.org/10.1007/s13201-022-01652-0)

Shaji E and Pradeepkumar AP. 2019. Surface Water and Groundwater Resources of Kerala: Management Issues, Policies, Future Strategies Surface Water and Groundwater Resources of Kerala: Management Issues, Policies, Future Strategies. Proceedings of the National Seminar jointly organized by Dept. of Geology, University of Kerala & Geological Society of India.

Taloor AK, Pir RA, Adimalla N, Ali S, Manhas DS, Roy S, Singh AK. Spring water quality and discharge assessment in the Basantar watershed of Jammu Himalaya using geographic information system (GIS) and water quality Index (WQI). Groundw Sustain Dev, 10, (2020):100364 (https://doi.org/10.1016/j.gsd.2020.100364)

Varma A. "Groundwater Resource and Governance in Kerala; Status, Issues and Prospects", 2017. (https://www.soppecom.org/pdf/Groundwater-Resource-and-Governance-in-Kerala.pdf)

4

Assessing Spatial Changes in Blue and Green Water Resources A Case Study of Valapattanam River Basin Using SWAT Model

***Sharannya T.M.*[1] *and Farissa Razack*[2]**

[1]*Hydrology and Climatology Research Group, Centre for Water Resources Development and Management (CWRDM), Kunnamangalam, Kozhikode, India*

[2]*Department of Environmental Science, Kannur University, Kannur, India*

Abstract

Climate change and human activities have significantly impacted water resources, leading to increased water scarcity in various regions. This study explores the spatial dynamics of blue and green water resources in the Valapattanam River Basin, Kerala amidst changing environmental conditions. Utilizing the Soil and Water Assessment Tool (SWAT) model and integrating with the Remote Sensing (RS) and Geographic Information Systems (GIS), the research assesses the impacts of climate variability and land use changes on water availability. Results indicate varying trends in precipitation, actual evapotranspiration, soil water content, surface runoff, and blue water across different sub-basins. The findings highlight the critical need for holistic water resource management strategies to ensure sustainable utilization in the context of changing climatic conditions.

Introduction

Climate change and human activities have emerged as significant factors affecting water resources in diverse regions (Kumar et al. 2022; Haddeland et al. 2014). The escalation in human water consumption for domestic and industrial purposes enhanced by the impacts of climate change has resulted in an overall decline in water availability (Liu et al. 2017; Paul et al. 2021). Rising temperatures and altered precipitation patterns directly influence water availability for ecosystems and agriculture potentially leading to greater evapotranspiration rates, intensified rainfall events, and prolonged droughts (Sharannya et al. 2021; Myers et al.

2023). Furthermore, urbanization, deforestation, and agricultural practices alter landscapes, affecting water resources in these areas.

Water shortages pose threats to ecosystem health, and socio-economic development and disrupt agricultural production (Piao et al. 2010). Thus, ensuring sustainable and efficient water usage necessitates accounting for water usage and quantifying it in a changing environment. To comprehensively assess the significance of water resources in the environment, Falkenmark (1995) introduced the concept of blue water (BW) and green water (GW) water, particularly for managing water resources in water-stressed regions. Green water refers to the portion of precipitation that infiltrates into the unsaturated soil layer and is subsequently utilized by plants through transpiration and soil evaporation. In fact, "blue water" represents surface and groundwater flow in rivers, lakes, and aquifers (Abate et al 2024; Sharannya et al. 2021). While numerous studies have investigated the spatial and temporal variation of BW and GW using the Soil and Water Assessment Tool (SWAT) (Feng et al. 2021; Li et al. 2009; Lyu et al. 2019), fewer publications have focused on assessing these variations and uncertainties at the basin scale under varying climate scenarios (Sharma et al. 2023). Hydrological models such as SWAT, serve as valuable tools for simulating past and future hydrological changes in the watersheds using projected climate data. SWAT specifically can simulate both blue and green flows at the watershed scale.

Though many studies have investigated trend analysis to evaluate the variability of runoff or surface water levels (i.e., BW flow), a limited number of works have concentrated on assessing these variations in spatial scale representation and evaluation under changing climatic conditions (Mao et al. 2018; Haddeland et al. 2014). Consequently, there exists an urgent necessity to thoroughly assess the spatio-temporal variations of both blue and green water and their interactions in response to a changing climate (Sharma et al. 2023; Zhu et al. 2018). This study focuses on analyzing the spatial variation in blue and green water in Valapattanam River Basin throughout the historical period, aiming to provide comprehensive insights into water resource dynamics and their responses to environmental changes.

Modeling Water Resources with SWAT: Assessing Blue and Green Water

The SWAT stands as a pivotal model in hydrological science, facilitating the evaluation of blue and green water components crucial for water resource management. By simulating various hydrological processes, SWAT enables the assessment of surface runoff, stream flow, and evapotranspiration among other parameters. Through its integration with Remote Sensing (RS) and

Geographic Information Systems (GIS), SWAT effectively quantifies both blue water (surface and groundwater) and green water (soil moisture utilized by vegetation). This comprehensive approach aids in understanding water availability dynamics, informing sustainable water usage strategies amidst evolving environmental conditions.

Model description and setup

The SWAT is a physically based semi-distributed framework to assess the impacts of both climate variability and human activities on water resources (Neitsch et al. 2011). It is a powerful hydrological model which integrates seamlessly with RS and GIS technologies. RS provides valuable data on land cover, soil moisture, and vegetation, which are essential inputs for SWAT. GIS facilitates the management, analysis, and visualization of spatial data and allows SWAT to utilize accurate geographical information for modeling various watershed processes. Together, these technologies enhance the precision and efficiency of hydrological modeling, enabling better-informed decisions in water resource management and environmental planning. This model excels in evaluating essential hydrological parameters such as surface runoff, streamflow, and baseflow, while also facilitating water quality analysis by assessing sediment load and nutrient flow (Opere and Okello, 2011). Its utility extends to investigating the impacts of climate change on hydrological processes, as evidenced by numerous studies across various river basins globally, including the Dharoi catchment of the Sabarmati River basin, the Western desert of Iraq, Rwanda, the Xinjiang River Basin, the Western Ghats of India, and many more (Sharma et al. 2023; Jabar et al. 2023; Umugwaneza et al. 2022; Xudong Zhang et al. 2022; Sharannya et al. 2018). Through its division of watershed into several sub-basins and further segmentation into Hydrological Response Units (HRUs) based on geographical characteristics such as digital elevation models, land use/land cover maps, and soil data, SWAT operates at a detailed scale to simulate water transportation effectively (Arnold et al. 1998) as given below.

$$SW_t = SW_0 + \sum t_i = 1 \ (R_{day} - Q_{surf} - E_a - W_{seep} - Q_{gw})$$

where SW_t is the final soil water content in mm; SW_0 is the initial soil water content in mm; t is the time (days); R_{day}, Q_{surf}, and E_a are the amount of precipitation, surface runoff, and evaporation on day i in mm, respectively; W_{seep} is the amount of water entering the vadose zone from the soil profile on day i (mm) and Q_{gw} is the amount of the return flows on day i (mm).

Blue water and Green water component in SWAT

Blue water and green water are two key components in hydrology that represent distinct aspects of the water cycle and play essential roles in ecosystem functioning and human activities. Blue water refers to surface water and groundwater including rivers, lakes, and aquifers which are accessible for various purposes such as drinking, irrigation, and industrial use (Mao et al. 2018). It represents the visible and measurable water bodies that directly contribute to streamflow and water availability. In contrast, green water refers to the moisture stored in the soil and utilized by plants through processes like transpiration and soil evaporation. Green water supports vegetation growth, regulates microclimate conditions, and sustains terrestrial ecosystems. Unlike blue water, green water is often not readily visible, as it remains within the soil-plant-atmosphere continuum, but it is vital for maintaining soil moisture levels, supporting agriculture, and influencing local hydrological processes (Sharma et al. 2023). Understanding the dynamics of blue water and green water is crucial for effective water resource management and environmental conservation. While blue water is directly accessible and heavily relied upon for human needs, green water plays a critical role in sustaining natural ecosystems and agricultural productivity (Zhu et al. 2018). Strategies for sustainable water use and management should consider both components, balancing the exploitation of blue water resources with the preservation of green water resources to ensure long-term water security and ecosystem health. Integrated approaches, such as hydrological modeling using tools like the SWAT, enable the assessment of both blue water and green water components, aiding in informed decision-making and resource allocation for water management practices. The output parameters used in the SWAT model to calculate the blue and green water are (SURQ,mm), lateral flow (LATQ,mm), and the groundwater contribution to streams (GWQ,mm) , (DA_RCHG,mm), and the summation of these parameters are considered as blue water. Whereas, the actual evapotranspiration (ET,mm) is quantified as green water flow (GWF), and soil water (SW) is taken as green water storage (GWS). GWF and GWS contribute to the total green water (GW). All these parameters were taken from the SWAT output database.

Case Study: Valapattanam River Basin

Valapattanam River is the longest river situated in the Kannur district of Northern Kerala, India spanning 110 km in length. The river originates from the Brahmagiri hills of Western Ghats in Karnataka at an altitude of 1350 m amsl. It extends from 11°49′32.76″N to 12°13′33.466″N latitudes and 75°19′1.441″E to 75°58′37.286″E longitude (Figure 4.1). Major tributaries of this river are Sreekantapuram, Bavalipuzha, Venipuzha, and Aralampuzha. The basin is nearly level surface near the coast, undulating in the midland and steep-sided hills and mountain in the east. The total area of the watershed covered for the study is 1784.23 sq. km. The elevation of the watershed ranges from 0 to 1715m and the only reservoir is Pazhassi reservoir which drains a catchment area of 1028 sq.km. The land use includes Deciduous Forest, Agricultural Land-Generic, Medium Density

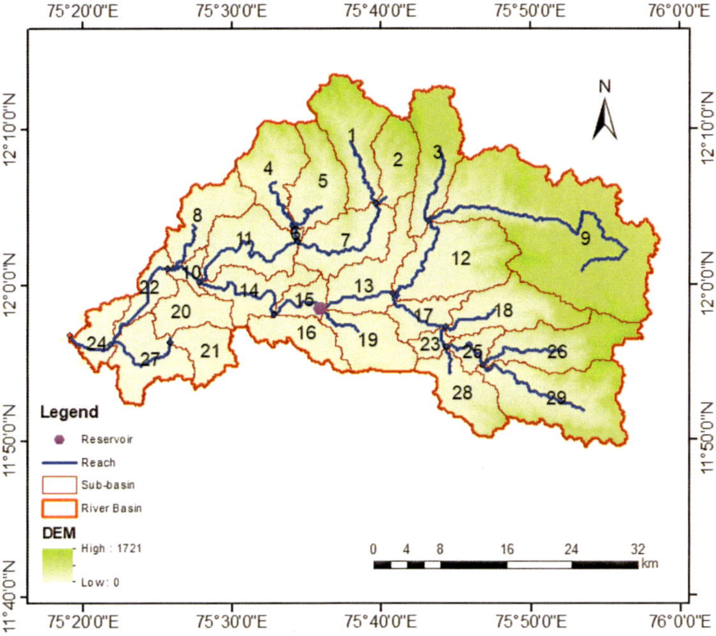

Fig. 4.1: The study area map

Residential area, Mixed Forest, Range-Grasses, Barren, Water, Coconut, and Evergreen Forests. The majority of the land comprises coconut *(more than 55% of the total area)* and the least being mixed forest *(less than 1%)*. Soil type mainly consists of Sandy Clay Loam *(more than 60%)*. The lower reaches of the Valapattanam River are extensively used for domestic and industrial activities. For the study purpose, the whole watershed was subdivided into 29 sub-basins and 239 HRU's (Hydrological Response Unit) by SWAT.

Hydrometeorology of the study area

The river basin experiences a predominantly hot and humid climate, with significant rainfall occurring during the Southwest monsoon (June to September) and the North-east monsoon (October to December). Approximately 70% of the total rainfall is recorded during the months of June to August. The catchment area receives an average annual rainfall of 3697 mm. The precipitation data was collected from four rain gauge stations (Kannur, Kottiyoor, Irikkur, and

Taliparamba) and the temperature data was collected from one station (Kannur) within the study area, sourced from the India Meteorological Department (IMD). Discharge data for the study was measured at the Perumannu gauging station maintained by the Central Water Commission, covering a period of 27 years from 1995 to 2022.

Land use and soil data of the study area

Land use of the study area is mainly classified into forest area, agriculture or plantation area, urban, grassland, barren, and water body. Coconut plantations constitute over 50% of the entire watershed, while the forest comprises evergreen, deciduous, and mixed types. Urban areas, accounting for 3% of the watershed, are primarily concentrated in the lower elevations. Barren land occupies the smallest portion, just 1% of the study area. Water bodies constitute approximately 2% of the total watershed, with additional areas covered by grasslands. The soil type of the Valapattanam watershed mostly consists of sandy clay loam (68%), followed by clay loam type C (22%), clay loam type D (6%), and 2% of sandy clay loam and sandy loam soil.

Model Calibration and Sensitivity Analysis

For the present study, ArcSWAT (version 2012) was run in ArcGIS10.4.1 platform to delineate watersheds. The delineated watersheds were subdivided into 29 sub-basins. Based on the land use and soil classification 239 HRUs were created by the model. For the weather data, daily precipitation data observed from four different weather stations was used and daily minimum and maximum temperature was also used to simulate the model. The model was calibrated using the observed discharge data from 1998 to 2015 leaving 3 years from 1995 to 1997 as warm-up periods. SWAT-CUP, SUFI2 algorithm which is a public domain computer program for calibration, validation, and uncertainty analysis of the SWAT model was used for this. Figures 4.2 and 4.3 illustrate the graphical representation of observed and simulated monthly streamflow data for both the calibration period (1988–2015) and the subsequent validation periods (2016–2022). It's worth noting that coefficients of determination (R^2) and Nash-Sutcliffe Efficiency (NSE) values for streamflow for calibration are 0.94 and 0.80 and the validation period is 0.88 and 0.78, respectively, while the Percentage Bias (PBIAS) values fall within ±20% for both calibration and validation periods. These results align with the criteria established by Moriasi et al. (2007) and demonstrate a commendable model performance.

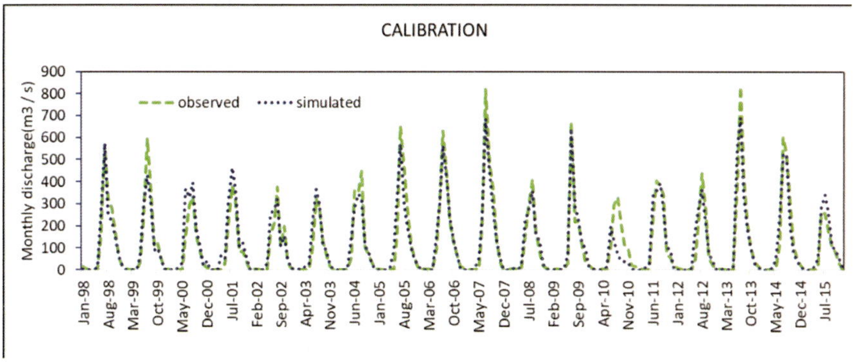

Fig. 4.2: Hydrograph during calibration period

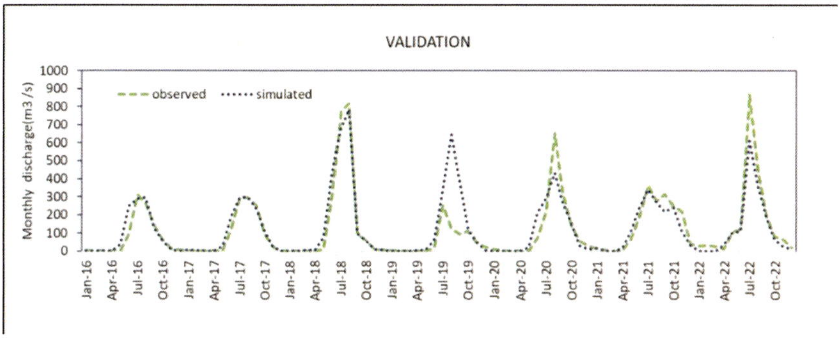

Fig. 4.3: Hydrograph during validation period

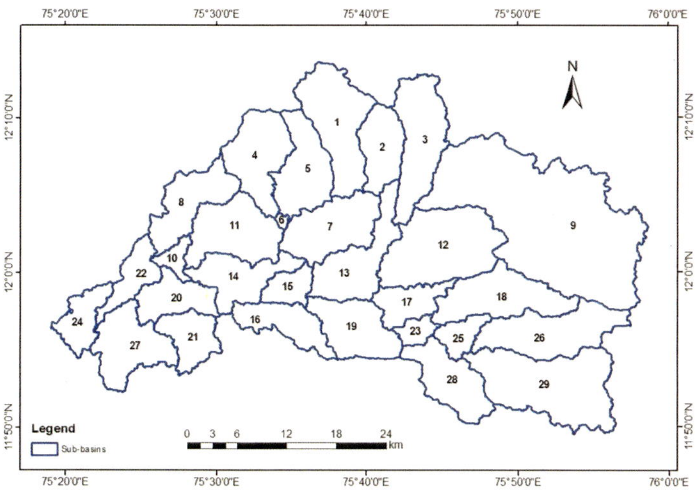

Fig. 4.4: Sub-basins of Valapattanam watershed

Subbasin representation of the study area

For the current study, Valapattanam watershed was further divided into 29 subbasins using SWAT (Figure 4.4). Each sub-basin was subdivided into several HRUs according to the geographical condition, including digital elevation model (DEM), land use/land cover map, soil map data, etc. The HRUs were the basic units that were used to calculate the transportation of water.

Spatial variation of rainfall

The study spanned over 25 years, i.e., from 1998 to 2022 which are divided into three time scales: 1998 to 2006 (T1), 2007 to 2014 (T2), and 2015 to 2022 (T3). Analysis of precipitation distribution (Figure 4.5) revealed a significant decrease in T1, despite an average annual precipitation of 3559 mm over the entire period. Sub-basins 9, 12, 18, 23, 25, 26, 28, and 29 consistently received higher precipitation, likely due to their proximity to the Western Ghats and higher elevation, with implications for water resource availability. Additionally, sub-basins 24 and 27, characterized by coconut plantations and lack of forest cover, exhibited consistent precipitation patterns across all three-time scales. The midlands showed an increasing trend in rainfall from T1 to T3, suggesting

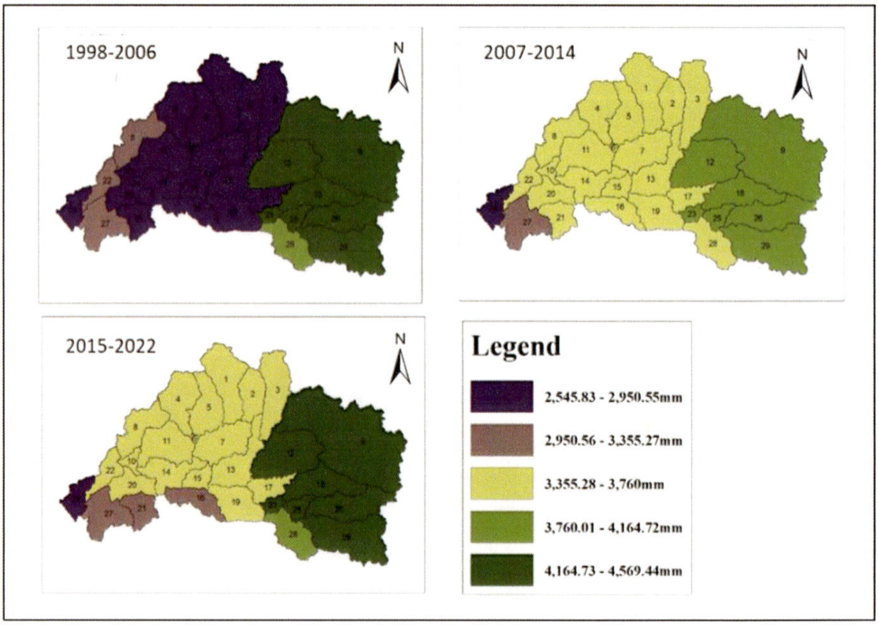

Fig. 4.5: Spatial representation of precipitation

potential shifts in local climate patterns. These findings underscore the spatial and temporal variability in precipitation distribution within the study area, highlighting the importance of considering local topography and land cover types in hydrological modeling and water resource management strategies.

Actual evapotranspiration

Actual Evapotranspiration (ETa), representing the water evaporated from surfaces and transpired by plants, increased consistently across all study periods. The spatial distribution of actual evapotranspiration depicted in Figure 4.6 revealed notable variations among sub-basins. Sub-basins 6 and 22 exhibited the highest Green Water Flow (GWF), while sub-basins 16 and 24 experienced the lowest. Sub-basin 6, characterized by extensive coconut plantations covering over 75% of its area, recorded the highest ETa levels. The observed ETa values ranged from 38.67 to 43.81 mm for the highest and 18.09 to 23.23 mm for the lowest. Interestingly, a trend of increasing ETa was observed from T2 to T3, indicating potential shifts in local hydrological processes over time. These findings have several implications for water resource management and agricultural practices in the study area. The significant increase suggests heightened water demand by vegetation that could affect

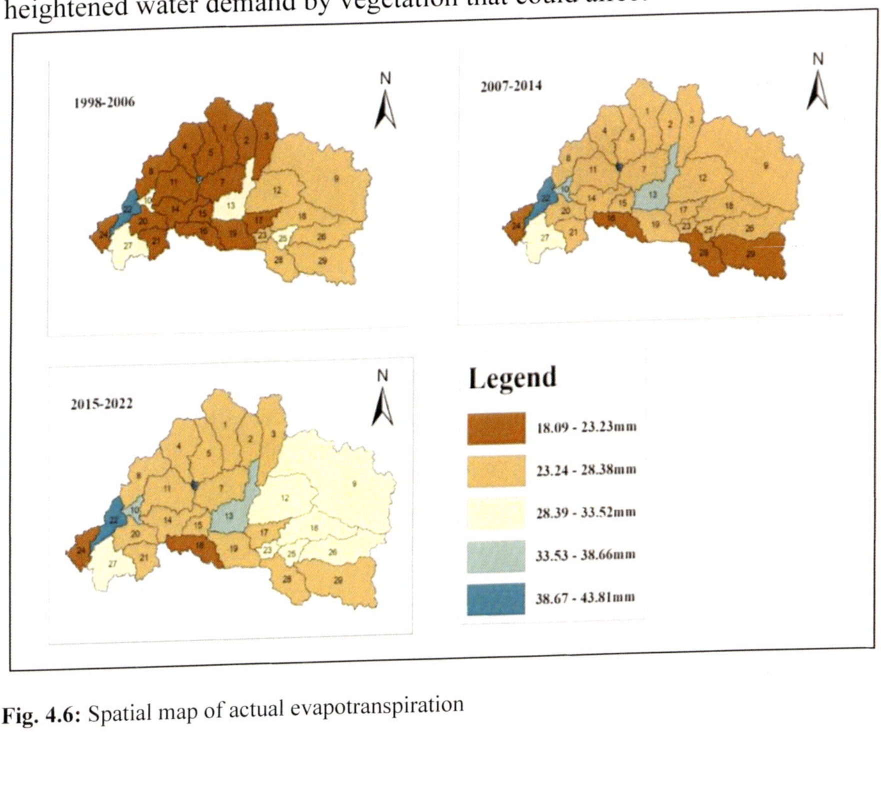

Fig. 4.6: Spatial map of actual evapotranspiration

local water availability and ecosystem dynamics. Sub-basins with high ETa levels, such as 6 and 22, may require targeted water management strategies to sustain agricultural productivity and ecosystem health. Conversely, sub-basins with lower ETa levels may indicate areas with less vegetation cover or limited water availability, necessitating conservation measures to mitigate potential environmental degradation. Additionally, the observed temporal trend of increasing ETa underscores the importance of considering long-term climate variations in water resource planning and management efforts.

Soil Water Content

Soil water content (SWC) serves as a critical hydrological parameter influenced by various factors including evapotranspiration, surface runoff, and water balance, playing a pivotal role in plant growth, soil temperature regulation, chemical transport, and groundwater recharge. Figure 4.7 illustrates a consistently increasing trend in SWC across the basin, ranging from 34 to 50.21 mm, with sub-basins 22 and 10 consistently exhibiting the highest values throughout the year. However, an alarming observation is the declining trend in SWC from period T2 to T3, particularly in the eastern part of the

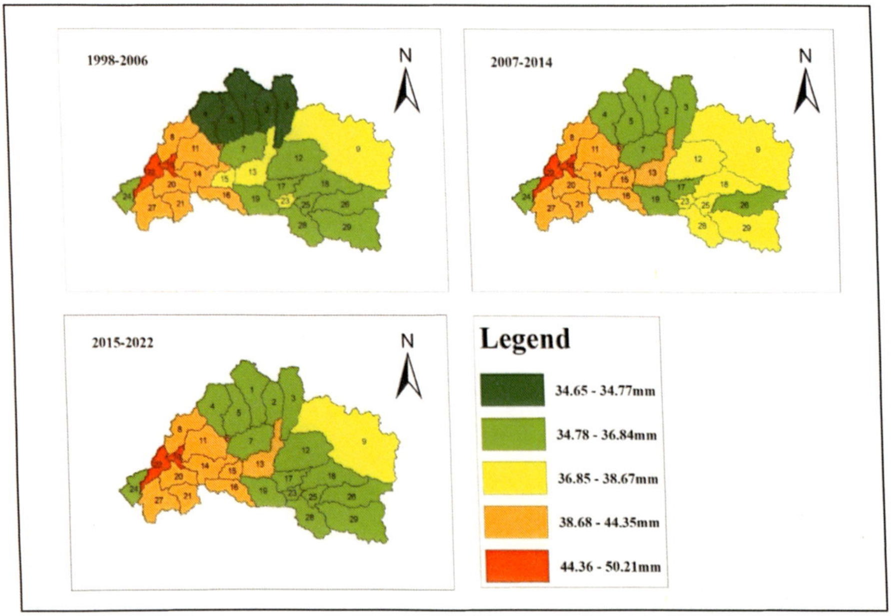

Fig. 4.7: Spatial map of soil water content

watershed. If this trend persists, it could lead to reduced moisture retention and adversely impact groundwater recharge in the region. These findings have significant implications for water resource management and ecosystem sustainability. The overall increase in SWC suggests favorable conditions for plant growth and ecosystem health across the basin. However, the declining trend in SWC in the eastern region highlights potential water stress and the need for targeted conservation measures to maintain soil moisture levels and support groundwater recharge. Strategies such as improved irrigation practices, afforestation, and land-use planning could help mitigate the adverse effects of declining SWC and ensure the sustainable utilization of water resources in the affected areas.

Surface Runoff Contribution

Surface runoff, driven by excess precipitation and limited soil infiltration capacity, exhibits spatial variability influenced by topographical characteristics and rainfall patterns. In the study area, higher rainfall contributes to increased runoff, particularly in regions with elevated elevation levels such as the eastern part of the watershed (Figure 4.8). Despite slight variations, the average surface runoff remains relatively consistent across the basin, highlighting stable

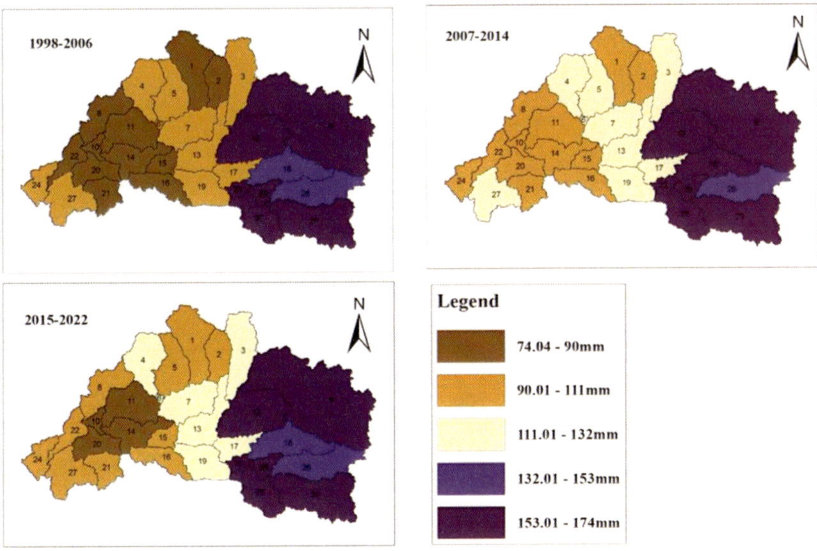

Fig. 4.8: Spatial map of surface runoff contribution

hydrological dynamics. However, notable trends include a decrease in runoff in the midland region during period T3, contrasting with an increase observed in period T2. Furthermore, a decreasing trend in surface runoff is noted along the coastal areas, coinciding with urban settlements. The observed increase in runoff in higher elevation regions underscores the importance of targeted measures to mitigate soil erosion and manage stormwater in these areas. Implementing green infrastructure practices such as permeable pavements and vegetated swales can help enhance infiltration and reduce runoff volumes, thereby minimizing the risk of flooding and preserving water quality. Moreover, the decreasing trend in coastal runoff suggests the need for sustainable urban drainage systems (SUDS) to manage runoff from impervious surfaces and alleviate pressure on drainage infrastructure. Integrating nature-based solutions and land-use planning strategies can promote resilience to climate change impacts and support the sustainable development of coastal communities.

Blue Water

Blue water, comprising surface and groundwater runoff, exhibits temporal variability influenced by hydrological processes and precipitation patterns. Blue water tends to increase after 2006 and decrease after 2015 except for the eastern part (Figure 4.9). The higher blue water content in the eastern

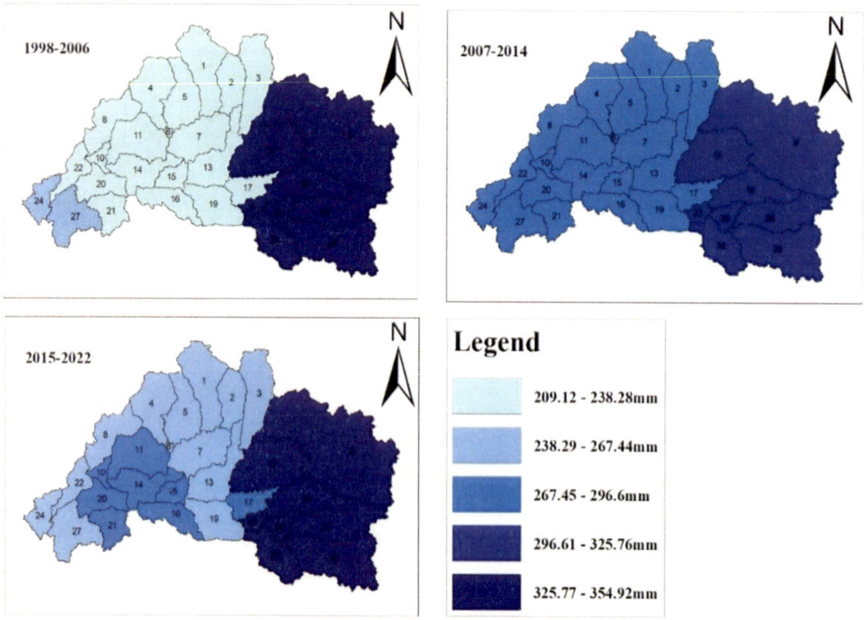

Fig. 4.9: Spatial map of blue water

sub-basins aligns with the precipitation distribution, highlighting the influence of rainfall on runoff generation. However, the lowest blue water levels during 1998–2006 and peak levels in T1 and T3 indicate significant fluctuations in water availability over time. The temporal variation in blue water highlights the need for adaptive management approaches that account for changing hydrological conditions. Implementing measures to enhance water retention and infiltration, such as soil conservation practices and wetland restoration, can help mitigate the impacts of fluctuating blue water levels on water security and ecosystem health. Additionally, promoting water-saving practices and efficient irrigation techniques can contribute to sustainable water use under variable runoff patterns. Continuous monitoring of blue water dynamics is essential for developing robust management strategies that enhance resilience to climate change and ensure the long-term sustainability of water resources.

Conclusions

The study provides valuable insights into the dynamics of blue and green water resources in the Valapattanam River Basin, under climate change, land use patterns, and water availability. Through the application of the SWAT model, integrated with RS and GIS, we have discerned significant spatial variations in key hydrological parameters across the basin. The findings reveal distinct trends in precipitation distribution, with certain sub-basins experiencing notable decreases in rainfall, particularly during the earlier period (T1), followed by subsequent increases or stabilization in later time scales (T2 and T3). Actual evapotranspiration rates exhibit a consistent rise over time, indicative of heightened water loss from the basin's surface and vegetation. Also, the soil water content demonstrates a decreasing trend from T2 to T3, reflecting a moisture retention decline and impacting the groundwater recharge and agricultural productivity. Surface runoff patterns exhibit spatial heterogeneity, with elevated runoff observed in areas characterized by higher precipitation and topographical features conducive to rapid runoff generation. The distribution of blue water resources underscores the influence of precipitation patterns, with sub-basins in the eastern region consistently exhibiting higher blue water availability compared to other areas.

Furthermore, the study emphasizes the critical importance of proactive water management strategies to mitigate the impacts of climate change and anthropogenic activities on water resources in the study area. Adaptive measures, informed by robust hydrological modeling and spatial analysis, are imperative to ensure the sustainable utilization of water resources in the Valapattanam River Basin and similar watersheds globally.

Acknowledgment

We acknowledge the Centre for Water Resources Development and Management (CWRDM) (Grant No. PP 94/22) dtd. March 8, 2023) for providing the fund and for their valuable support to the research.

References

Abate, B. Z., Alaminie, A. A., Assefa, T. T., Tigabu, T. B., and He, L. Modeling climate change impacts on blue and green water of the Kobo-Golina River in data-scarce upper Danakil basin, Ethiopia. Journal of Hydrology: Regional Studies, 53, (2024). https://doi.org/10.1016/j.ejrh.2024.101756

Arnold, J. G., Srinivasan, R., Muttiah, R. S., and Williams, J. R. Large Area Hydrologic Modeling and Assessment Part I: Model Development. Journal of The American water resources Association American Water Resources Association, 34, no. 1, (1998): 73-89.

Falkenmark, M., 1995. Coping with water scarcity under rapid population growth. In: Conference of SADC Ministers, Pretoria, vol. 23, p. 24.

Feng, C., Yang, L., and Han, L. Impacts of climate change on blue and green water resources in the Xiangjiang River Basin of the Yangtze River, China. Frontiers in Earth Science, 9, (2021): 677191.

Haddeland I, Heinke J, Biemans H, Eisner S, *et al.* Global water resources affected by human interventions and climate change. Proc Natl Acad Sci U S A. 111, no. 9, (2014): 3251-6. doi: 10.1073/pnas.1222475110. Epub 2013 Dec 16. PMID: 24344275; PMCID: PMC3948259.

Jabar, B. S., Sayl, K. N., and Alboresha, R. Incorporating a GIS-Based Approach and SWAT Model to Estimate Sediment in the Western Desert of Iraq. International Journal of Design and Nature and Ecodynamics, 18, no. 4, (2023): 791-800. https://doi.org/10.18280/ijdne.180405

Kumar, M., Denis, D. M., Kundu, A., Joshi, N., and Suryavanshi, S. Understanding land use/land cover and climate change impacts on hydrological components of Usri watershed, India. Applied Water Science, 12, no. 3, (2022). https://doi.org/10.1007/s13201-021-01547-6

Li, Z., Liu, W. Z., Zhang, X. C., and Zheng, F. L. Impacts of land use change and climate variability on hydrology in an agricultural catchment on the Loess Plateau of China. Journal of Hydrology, 377, no. 1-2, (2009): 35-42.

Liu, J., Yang, H., Gosling, S. N., Kummu, M., *et al.* Water scarcity assessments in the past, present, and future. Earth's Future, 5, no. 6, (2017): 545-59. https://doi.org/10.1002/2016EF000518

Lyu, L., Wang, X., Sun, C., Ren, T., and Zheng, D. Quantifying the effect of land use change and climate variability on Green water resources in the Xihe River Basin, Northeast China. Sustainability, 11, no. 2, (2019): 338.

Mao, G., Liu, J., Han, F., Meng, Y., Tian, Y., Zheng, Y., and Zheng, C. 2018. Assessing Green and Blue Water: Understanding Interactions and Making Balance between Human and Nature. Hydrology and Earth System Sciences Discussions, May, 1–26. https://doi.org/10.5194/hess-2018-193

Moriasi, D.N., *et al.* Model evaluation guidelines for systematic quantification of accuracy in watershed simulations. Tran. ASABE, 50 no. 3, (2007): 885–900. doi:10.13031/2013.23153

Mudbhatkal, A., & Amai, M. (2018). Regional climate trends and topographic influence over the Western Ghat catchments of India. International Journal of Climatology, 38(5), 2265–2279. https://doi.org/10.1002/joc.5333

Myers, D. T., Ficklin, D. L., and Robeson, S. M. Hydrologic implications of projected changes in rain-on-snow melt for Great Lakes Basin watersheds. Hydrology and Earth System Sciences, 27, no. 9, (2023): 1755-70. https://doi.org/10.5194/hess-27-1755-2023

Neitsch, S.L., Arnold, J.G., Kiniry, J.R., and Williams, J. R. 2011. S Oil and Water Assessment Tool Documentation.

Opere, A. O., and Okello, B. N. (2011). Hydrologic analysis for river Nyando using SWAT. Hydrology and Earth System Sciences Discussions, 8, no. 1, (2011): 1765-97. https://doi.org/10.5194/hessd-8-1765-2011

Paul, M., Rajib, A., Negahban-Azar, M., Shirmohammadi, A., and Srivastava, P. Improved agricultural Water management in data-scarce semi-arid watersheds: Value of integrating remotely sensed leaf area index in hydrological modeling. Science of the Total Environment, 791, (2021). https://doi.org/10.1016/j.scitotenv.2021.148177

Piao, S., Ciais, P., Huang, Y., *et al.* The impacts of climate change on water resources and agriculture in China. Nature, 467, no. 7311, (2010): 43-51. https://doi.org/ 10.1038/nature09364.

Sharannya T M., Kolluru, V., Amai, M., and Acharya, T. D. Enhanced streamflow simulations using nudging based optimization coupled with data-driven and hydrological models. Journal of Hydrology: Regional Studies, 43, (2022). https://doi.org/10.1016/j.ejrh.2022.101190

Sharannya, T. M., Mudbhatkal, A., and Mahesha, A. Assessing climate change impacts on river hydrology – A case study in the Western Ghats of India. Journal of Earth System Science, 127, no. 6. (2018). https://doi.org/10.1007/s12040-018-0979-3

Sharannya, T. M., Venkatesh, K., Mudbhatkal, A., Dineshkumar, M., and Mahesha, A. Effects of land use and climate change on water scarcity in rivers of the Western Ghats of India. Environmental Monitoring and Assessment, 193, no. 12, (2021). https://doi.org/10.1007/s10661-021-09598-7

Sharma, A., Patel, P. L., and Sharma, P. J. Blue and green water accounting for climate change adaptation in a water scarce river basin. Journal of Cleaner Production, 426, (2023). https://doi.org/10.1016/j.jclepro.2023.139206

Umugwaneza, A., Chen, X., Liu, T., *et al.* Integrating a GIS-based approach and a SWAT model to identify potential suitable sites for rainwater harvesting in Rwanda. Aqua Water Infrastructure, Ecosystems and Society, 71, no. 4, (2022): 415-32. https://doi.org/10.2166/aqua.2022.111

Zhao, A., Zhu, X., Liu, X., Pan, Y., and Zuo, D. Impacts of land use change and climate variability on green and blue water resources in the Weihe River Basin of northwest China. Catena, 137, (2016): 318-27. https://doi.org/10.1016/j.catena.2015.09.018

5

GIS Tools for Mapping the Spatio-Temporal Trends in Rainfall Distribution A Case Study of Kerala State, India

Fathima J.P., Shahanas P.V., Naveena K., and Naga Kumar K. Ch. V.

Centre for Water Resources Development and Management, Kunnamangalam Kozhikode, India

Abstract

Understanding the spatial distribution of rainfall is crucial for effective management of water resources and agriculture. In this study, we have analyzed the rainfall data at the district level for the state of Kerala, a region in Southern India to demonstrate the potential of Geographic Information System (GIS) tools toward mapping the non-spatial data i.e., rainfall data and understanding its spatial distribution. In this study, we have used the GIS tools to visualize the rainfall distribution at the district level using the interpolation technique, i.e., Inverse Distance Weighting (IDW). The results indicate that the northern districts of Kerala have experienced higher rainfall levels during the Southwest monsoon and throughout the year whereas the southern districts are characterized by increased rainfall during the summer, Northeast monsoon, and winter periods.

Keywords: *Geographic Information System (GIS), Inverse Distance Weighted (IDW), Kerala state (India), Rainfall data, Spatial distribution*

Introduction

Rainfall is one of the climatic variables that varies very highly spatially and temporally. Understanding this micro-level variation is a crucial indicator for planning future water requirements and preparing for management strategies. Numerous researchers have delved into the distribution, variability, and trends of rainfall across diverse spatial scales, ranging from global to regional and basin levels. Nair et al. (2014) analyzed the rainfall trend and seasonality index for Kerala state in India over the last 100 years. The rainfall trend was found to be decreasing significantly in almost all the regions of the state. Nischitha et al. (2013) provide insights into the spatial and temporal variability of daily monsoon rainfall in the Tunga and Bhadra River basins in Karnataka state. Paradkar et al. (2019) evaluated the pattern of weekly rainfall by using statistical parameters such as mean, standard deviation, skewness, and kurtosis for five stations in Pratapgarh district.

Geographical Information System (GIS) proved to be an effective tool in handling information and analyzing rainfall patterns, providing visual representations, and facilitating the decision-making process (Kadir et al. (2016) and Rao et al. (2017)). Anand et al. (2020) analyzed the long-term spatial and temporal trends of rainfall in the Lower Bhavani basin in Tamil Nadu, India using GIS and statistical methods. Earls et al. (2007) compared different interpolation methods such as Inverse Distance Weighting, Spline, and Kriging (ordinary and universal) to generate a 30 m representation of the rainfall data. Spatial interpolation techniques perform differently and their evaluation should not rely solely on quantitative assessment. Wu & Hung (2016) compared the performance of three spatial interpolation techniques i.e., IDW, kriging, and spline using quantitative assessment, 2D visualization, and 3D visualization. IDW and kriging performed similarly in terms of quantitative assessment, while spline performed worse. IDW and spline performed similarly in 2D visualization, while kriging showed different behavior. Kriging was observed to be an inexact interpolation in 3D visualization, while spline tended to create extreme values along the edges of the study area. Some studies discovered that the accuracy of Kriging in the study area is higher than that of IDW or Thiessen interpolations (Earls et al. 2007; Song et al. 2008; Chen et al. 2010). GIS applications have the capability to integrate spatial data and non-spatial data like rainfall and to visualize, model, and analyze how rainfall distribution varies across space and over time, providing insights into climate change, water resource management, agricultural planning, and disaster management. The present study aims to analyze annual rainfall data pertaining to Kerala state in the GIS environment.

Methodology

GIS is used to visualize the spatial pattern of rainfall across the administrative boundaries (district) of Kerala state. The spatial distribution of rainfall was studied with the help of the interpolation technique, i.e., Inverse Distance Weighting (IDW), and respective maps were prepared in a GIS environment. The method of interpolation using GIS typically involves several steps to estimate values at unmeasured locations within a study area based on known data points. The methodology for spatial analysis begins with data collection, where spatial data, including known data points and the boundary of the study area, are gathered. This collection phase often involves acquiring rainfall data from rain gauge stations along with the locations of each station and any other pertinent spatial information. Following this, the data preparation phase ensures that all collected data is formatted suitably for GIS analysis, which may entail processes such as geo-referencing, projection matching, and cleansing the data of any errors or outliers. Next, in the interpolation process, the chosen interpolation method is applied to the known data points within the study area. This step includes executing the interpolation tool available in the GIS software, where specific parameters like the interpolation method, search radius, and output resolution are set. The final phase involves visualization and interpretation, where the interpolated surface is visualized using GIS software to produce geo-spatial maps or other visual representations that depict the estimated values across the study area. Interpreting these results yields insights into the spatial distribution of the variable of interest, providing a comprehensive understanding of the analyzed spatial data.

Case Study

The daily rainfall data for all the rain gauge stations in Kerala state spanning a period of 52 years (1970–2022) was obtained from the India Meteorological Department (IMD). This data was then utilized for spatial-temporal mapping of rainfall across Kerala. Subsequently, the chronologically arranged data was re-modified on a monthly basis, as well as aggregated seasonally into Summer (March, April, May), South-west monsoon (June, July, August, and September), North-east monsoon (October, November), Winter (December, January, and February), and annually. The data processing of rainfall data for various districts of Kerala was performed using Microsoft Office. Following this, trend analysis was conducted utilizing the modified Mann-Kendall test, employing the "modifiedmk" package within the R software environment. Figure 5.1 shows the location map of the study area and trend analysis results are given in Tables 5.1 and 5.2.

Trend Analysis Of Spatio-Temporal Movement of *Rainfall Pattern Across Kerala (1970–2022)*

The rainfall pattern in Kerala state showed notable variations during specific months and seasons in different districts. The results of the trend analysis conducted on monthly, seasonal, and annual rainfall data, focusing on their spatial distribution are given below.

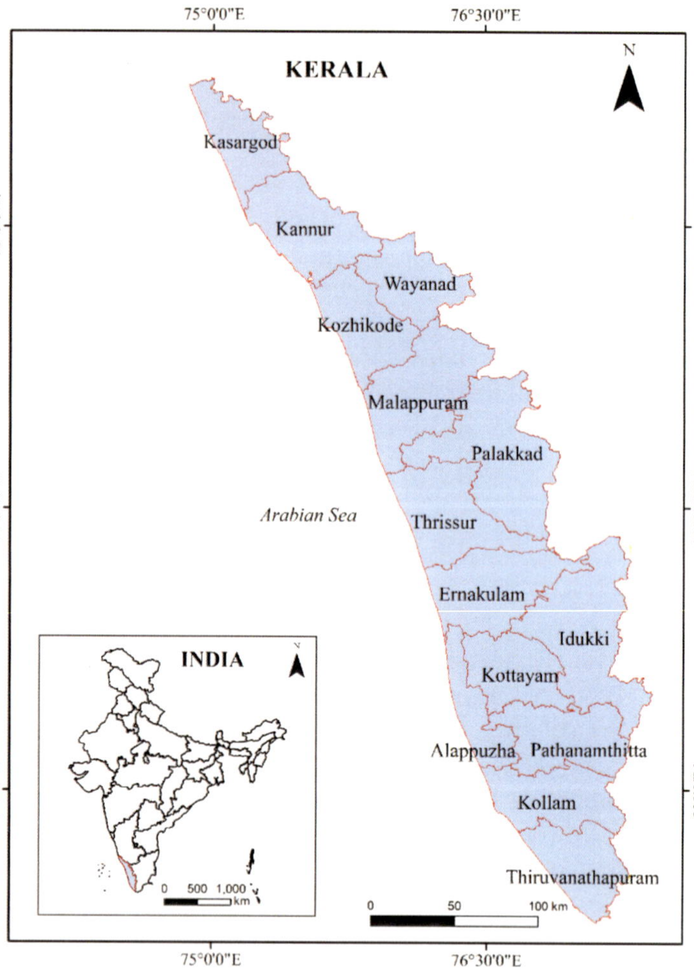

Fig. 5.1: Location map of the study area

Monthly rainfall trend

Figure 5.2 illustrates the trends in monthly rainfall across districts in the state of Kerala from 1970 to 2022. Some districts display specific months with significant increasing or decreasing trends in rainfall. For instance, Alappuzha generally maintains stable rainfall patterns throughout the year, except for a slight increase in March. In Thrissur, there is an increasing trend in rainfall observed in August. In districts like Kozhikode and Kasaragod, rainfall patterns remain stable throughout the year, except for September, which exhibits an increase in rainfall. Moreover, Idukki demonstrates notable variations in

Table 5.1: Modified Mann-Kendall Test Z-Values for Monthly Rainfall Data of Kozhikode

Districts	**Jan**	**Feb**	**Mar**	**Apr**	**May**	**Jun**	**Jul**	**Aug**	**Sep**	**Oct**	**Nov**	**Dec**
Trivandrum	−0.29	−0.99	−0.3	−0.36	0.45	−0.21	−0.79	−0.18	0.37	1.1	0.65	0.52
Kollam	1.02	0.23	1.61	0.22	0.57	−1.42	−2.14*	0.27	**1.72**	1.15	0.3	1.02
Pathanamthitta	0.57	−1.43	1	0.07	−0.46	−1.64	−2.57*	0.29	1.32	−1.5	0.04	1.39
Alappuzha	0.28	−0.69	**1.84**	−0.72	0.22	−0.69	−1.61	−0.72	−0.02	0	−1.04	−0.02
Kottayam	0.82	−0.68	**1.72**	1.04	0.81	−2.54*	−0.3	0.11	0.7	1.25	−0.53	0.45
Idukki	0	−1.04	**1.93**	−0.04	0.54	−1.21	−2.28*	1.96*	0.89	−2.21*	−0.86	1.57
Ernakulam	0.12	0.99	1.57	1.99*	0.67	−1.44	−1.62	−0.43	1.1	1.27	−0.77	0.07
Thrissur	−0.47	−0.66	0.08	−0.08	−0.15	−0.15	0	1.52	0	-0.23	0.23	0.53
Palakkad	−0.38	0.06	1.1	−1.01	0.98	0.77	−0.92	1.1	1.13	**−1.69**	0	0.34
Malappuram	−0.83	0.12	0.57	−0.82	−1.17	−1.17	−0.07	**1.66**	−0.07	**−1.81**	−0.52	0.32
Kozhikode	0	0.74	1.47	0.59	−0.17	−1.27	−1.32	−1.02	**1.76**	1.29	−0.62	0.21
Wayanad	0.08	−0.38	2.42*	0.39	−0.24	0.32	−0.58	2.27*	2.53*	−0.96	−0.02	1.14
Kannur	−0.23	0.39	1.33	1.28	0.8	**−1.65**	0.39	−0.27	2.32*	1.98*	0	0.29
Kasargod	0.81	0.45	1.3	0.6	−0.24	−1.14	−1.48	1.11	**1.89**	−0.39	0.21	1.41

Table 5.2: Modified Mann-Kendall Test Z-Values for Annual and Seasonal Rainfall Data

Districts	Annual	Summer	South West Monsoon	North-East Monsoon	Winter
Trivandrum	0.67	0.2	0.3	1.19	0.18
Kollam	0.9	1.21	−0.05	0.84	−0.15
Pathanamthitta	**−1.89**	0.25	−1.53	−1.53	−0.32
Aalappuzha	−0.97	0.13	−1.44	−0.65	−0.59
Kottayam	0.42	**1.67**	−0.64	0.95	−0.13
Idukki	−0.39	1	−0.57	−2.11*	0.46
Ernakulam	0.32	1.22	−0.35	0.47	0.6
Thrissur	0	−0.68	0.61	0.38	−1.44
Palakkad	0.56	0.15	1.22	−1.33	−0.27
Malappuram	−0.57	−1.46	0.62	**−1.91**	0
Kozhikode	−0.89	0	−0.85	0.25	0.42
Wayanad	1.41	0.62	1.41	−0.62	−0.09
Kannur	0.87	0.65	0.46	1.96*	0.07
Kasargod	−0.81	−0.28	−0.43	−0.99	1.56

Bold: 10% statistical significance; * indicates 5% statistical significance

rainfall during March, July, August, and October. The months showing more fluctuation in rainfall patterns across the districts in Kerala can be identified as March, June, July, August, and September. In March, there was a positive trend in rainfall across the Kerala state, except in Thiruvananthapuram district, which experienced a negative trend, though it is not significant. In June, there is a general trend of decreasing rainfall, with Kannur and Kottayam experiencing notable declines. Only Wayanad and Palakkad saw increases in rainfall, though these were not significant. This pattern clearly indicates a reduction in rainfall during June. July month follows a similar pattern to June, where only Kannur district exhibits an increasing trend in rainfall. Idukki, Pathanamthitta, and Kollam displayed significant declines in rainfall, whereas Thrissur district showed no trend during July. In August and September months, it exhibited an increasing trend in rainfall. In August, only 5 out of 14 districts showed a decreasing trend, which is not significant. Among the districts with an increasing trend, Wayanad, Malappuram, and Idukki show a significant increase in rainfall. In September, all districts except Malappuram and Alappuzha, which experienced a non-significant decrease in rainfall, showed an increasing trend. Notably, northern Kerala, including Kasargod, Kannur, Kozhikode, and Wayanad districts displays a significant increase in rainfall. Thrissur, however, shows no visible trend.

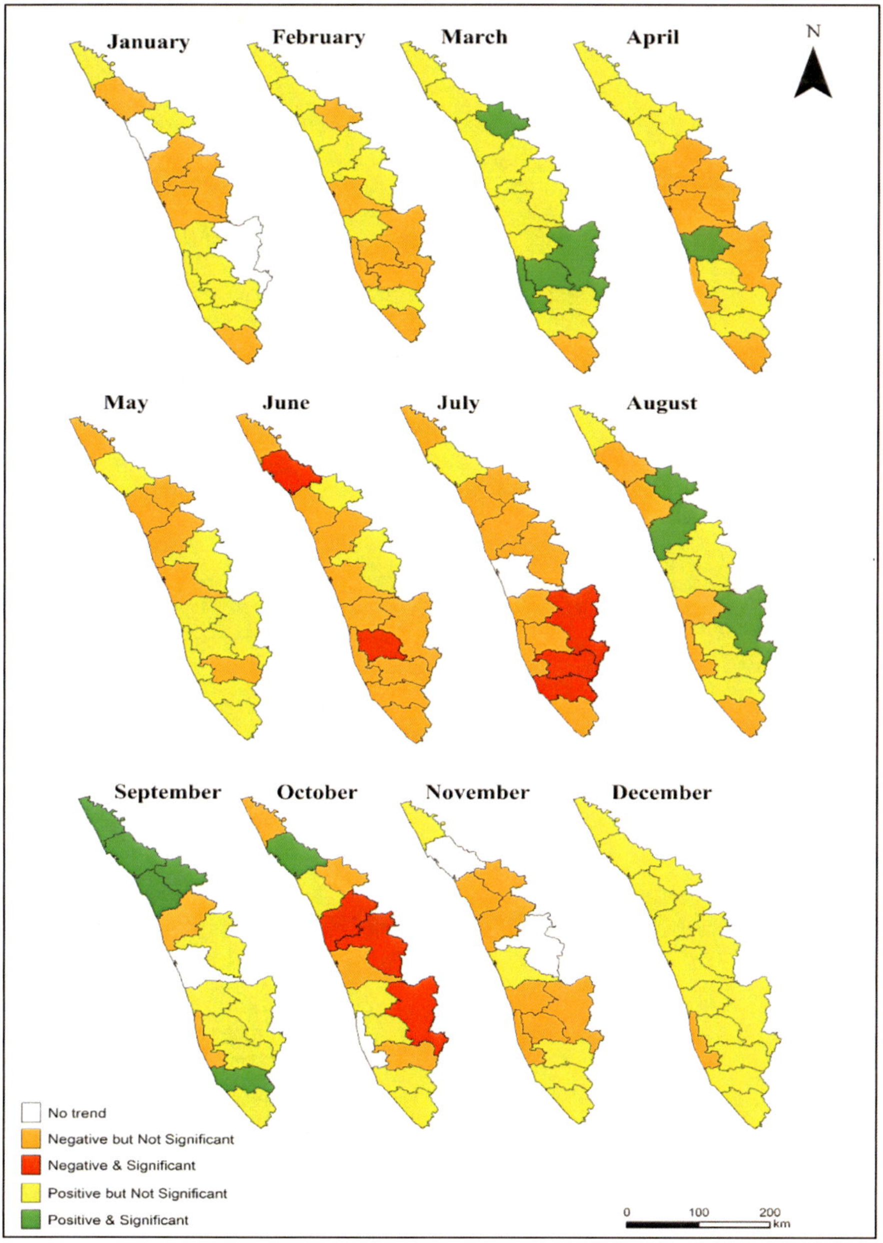

Fig. 5.2: District-wise trends in the monthly rainfall over the Kerala state (1970–2022).

Seasonal rainfall trend

Figure 5.3 depicts the trends in seasonal rainfall across the districts in the Kerala state from 1970 to 2022. Over the past 52 years, both Kannur and Thiruvananthapuram have displayed a consistent, though not significant increasing trend in rainfall across all seasons. During the Southwest monsoon period, southern Kerala has received more rainfall than its northern counterparts, with the exceptions of Kasargod and Kozhikode, both of which have demonstrated a decreasing trend in rainfall. The Northeast monsoon season has shown more variability; Kannur has experienced a significantly increasing trend in rainfall whereas Idukki has seen a significantly decreasing trend, highlighting the diverse impact of monsoonal changes across the Kerala region.

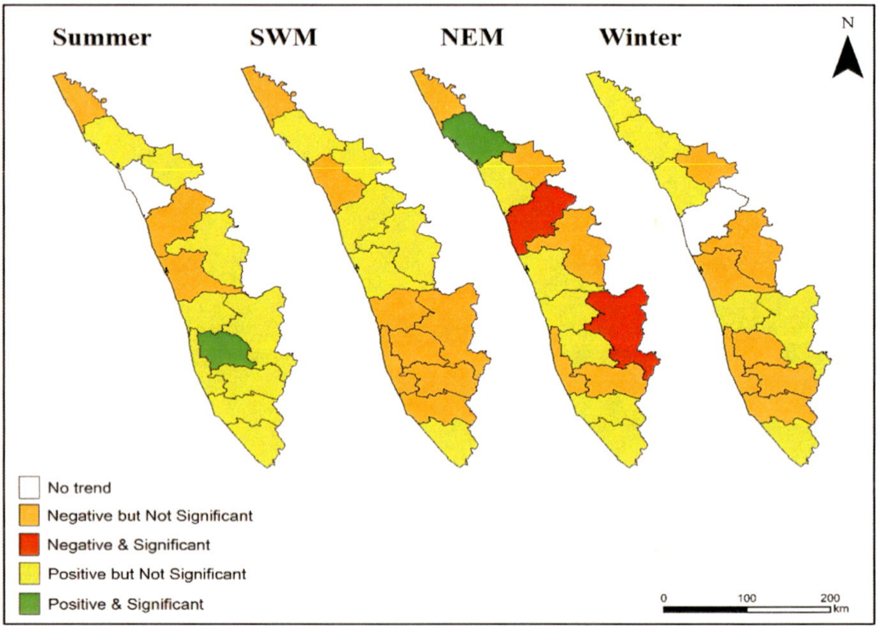

Fig. 5.3: District-wise trends in the seasonal rainfall over the Kerala state (1970–2022)

Annual rainfall trend

District-wise trends in the annual rainfall over the Kerala state are given in Figure 5.4. The annual rainfall trend across Kerala indicates a general lack of significant change state-wide. However, specific districts such as Kasargod, Kozhikode, Malappuram, Idukki, Alappuzha, and Pathanamthitta districts are experiencing a decreasing trend in rainfall. Conversely, Thrissur stands out for its stable rainfall pattern, showing no trend. Meanwhile, the remaining districts have shown an increasing trend in annual rainfall, showcasing varied climatic behavior across different regions of the state.

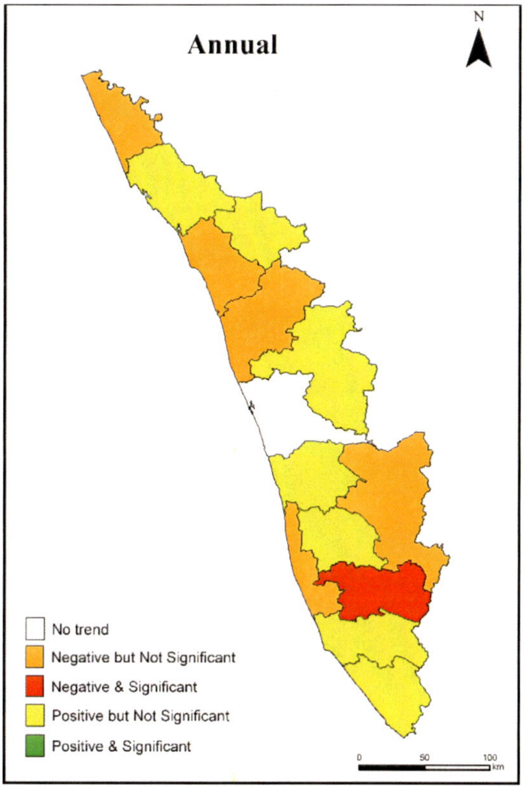

Fig. 5.4: District-wise trends in the annual rainfall over the Kerala state (1970–2022)

Conclusions

The interpolation tool (IDW) in the GIS environment proved valuable in mapping the spatial distribution of rainfall data over the Kerala state, India. Through rainfall map preparation, the study has demonstrated the spatial variations of rainfall across various districts in Kerala state, facilitating the identification of areas characterized by both high and low rainfall. GIS helps in modeling these spatial variations into an effortlessly interpretable frame that can be used for other related purposes. The present study focused on analyzing annual rainfall patterns within the study area using GIS techniques, and also toward demonstrating the efficacy of GIS in understanding the rainfall variability for the researcher, decision-making, etc.

References

Anand, B., Karunanidhi, D., Anand, B., Karunanidhi, D., Anand, B., & Karunanidhi, D. Long term spatial and temporal rainfall trend analysis using GIS and statistical methods in Lower Bhavani basin, Tamil Nadu, India. Indian Journal of Geo-Marine Sciences, 49, no. 3, (2020): 419-27.

Chen, D., Ou, T., Gong, L., Xu, C. Y., Li, W., Ho, C. H., & Qian, W. Spatial interpolation of daily precipitation in China: 1951–2005. Advances in Atmospheric Sciences, 27, (2010):1221-32.

Childs, C. Interpolating surfaces in ArcGIS spatial analyst. ArcUser, July-September, 3235, no. 569, (2004): 32-5.

Earls, J., & Dixon, B. 2007. Spatial interpolation of rainfall data using ArcGIS: A comparative study. In Proceedings of the 27th Annual ESRI International User Conference, Vol. 31, pp. 1-9.

Kadir, A. A., Kaamin, M., Azizan, *et al.* 2016, July. Application of Geographic Information Systems (GIS) in Analysing Rainfall Distribution Patterns in Batu Pahat District. In IOP Conference Series: Materials Science and Engineering, Vol. 136, No. 1, p. 012064. IOP Publishing.

Nair, A., Joseph, K. A., & Nair, K. S. Spatio-temporal analysis of rainfall trends over a maritime state (Kerala) of India during the last 100 years. Atmospheric Environment, 88, (2014): 123-32.

Nischitha, V., Ahmed, S. A., Varikoden, H., Revadekar, J. V., & Srinivasa Reddy, G. S. Spatial and temporal variability of daily monsoon rainfall in Tunga and Bhadra River basins, Karnataka. Annals of GIS, 19, no. 4, (2013): 219-30.

Paradkar V.D., Mittal, H. K., Singh, P. K., Kothari, M., & Jain, H. K. Analysis of Spatial and Temporal Rainfall Variability of Pratapgarh, India. Ind. J. Pure App. Biosci, 7, no. 5, (2019): 36-42.

Rao, G. S., Giridhar, M. V. S. S., & Mohan, S. 2017. Rainfall analysis with reference to Spatial and Temporal in Kaddam Watershed Using Geomatics. In National Conference on Water Environment and Society, p. 540.

Song, L., Tian, Y., & Wu, L. (2008). On comparison of spatial interpolation methods of daily rainfall data: a case study of Shenzhen. Geo-Inf. Sci, 10, (2018): 566-72.

Wu, Y. H., & Hung, M. C. Comparison of spatial interpolation techniques using visualization and quantitative assessment. Applications of spatial statistics, (2016): 17-34.

Xu WenBo, X. W., Zou YangJuan, Z. Y., Zhang GuoPing, Z. G., & Linderman, M. 2015. A comparison among spatial interpolation techniques for daily rainfall data in Sichuan Province, China.

6

Geographic Information System (GIS): A Practical Guide Towards Data Visualization, Analysis and Map Preparation

Naga Kumar K. Ch. V.[1], Nasih K[2], Thomas N.[3], Namitha T[4] and Naveena K.[4]

[1–4]Centre for Water Resources Development and Management (CWRDM), Kozhikode Kerala, India

Introduction to GIS

QGIS (Quantum GIS) is an open-source Geographic Information System (GIS) software has been used to showcase various exercises pertaining to viewing, editing, analyzing, and printing of the geospatial data. QGIS helps to allow the users to easily access and manage geographic data stored in folders on local disks or relational databases available on the user's network. QGIS supports raster, vector, and mesh layers. QGIS also allows using data from external sources with the help of web services such as Web Map Service (WMS) and Web Feature Service. QGIS is a constantly developing software as anyone can add new features or update existing tools. Newer versions of QGIS are frequently released on the official website. It is a cross-platform software that can be installed in Windows, Linux, and MacOS.

Installation of Q GIS

1. Go to the source link

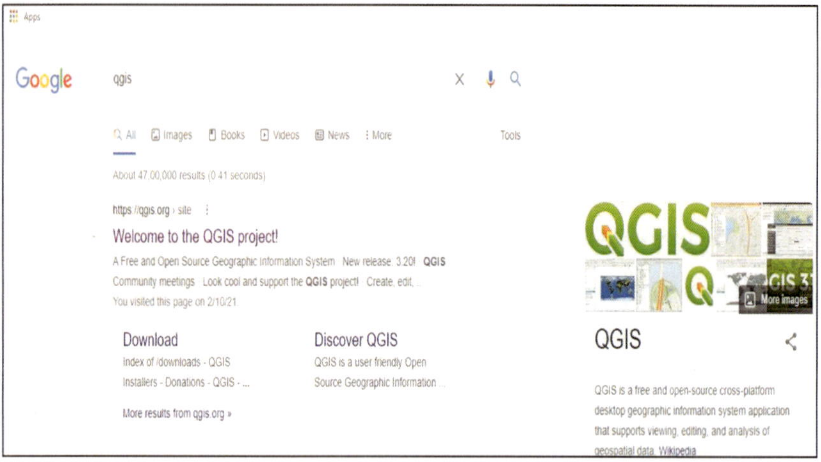

Source link: download.qgis.org/

2. Download QGIS Setup file

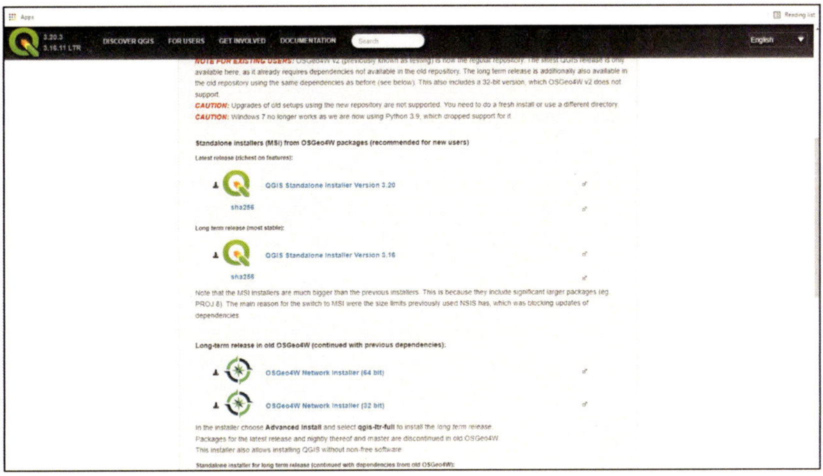

3. Find out setup file → Right click "Install"

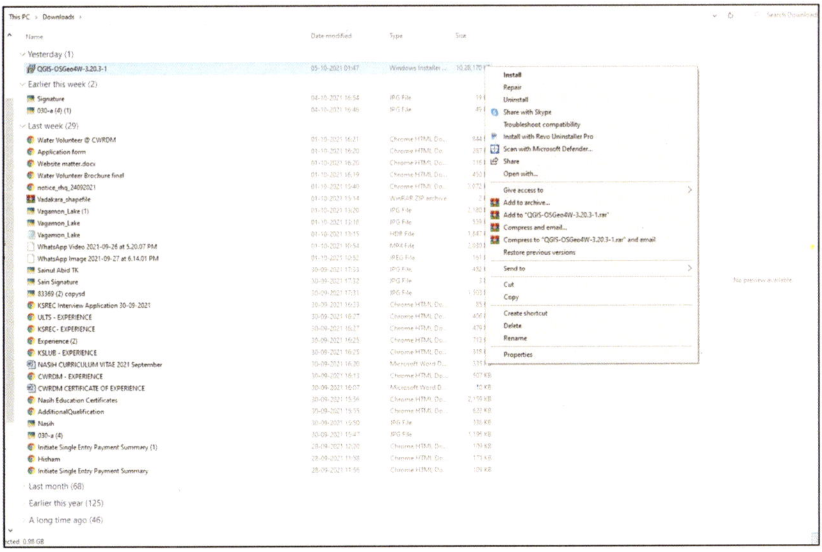

4. Run the setup click on "I Agree" → it will take time to Finish

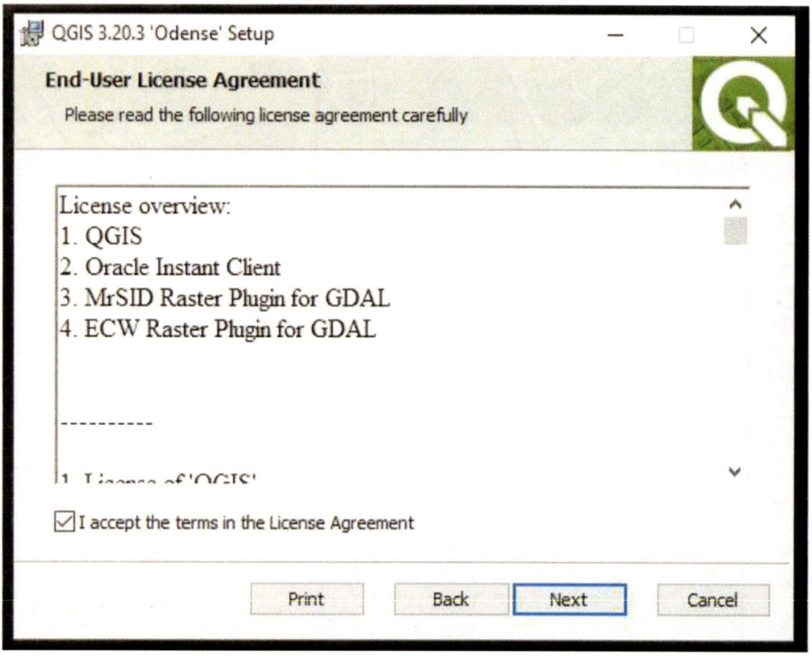
QGIS 3.20.3 'Odense' Setup
End-User License Agreement
Please read the following license agreement carefully
License overview:
1. QGIS
2. Oracle Instant Client
3. MrSID Raster Plugin for GDAL
4. ECW Raster Plugin for GDAL

I accept the terms in the License Agreement
Print
Back
Next
Cancel

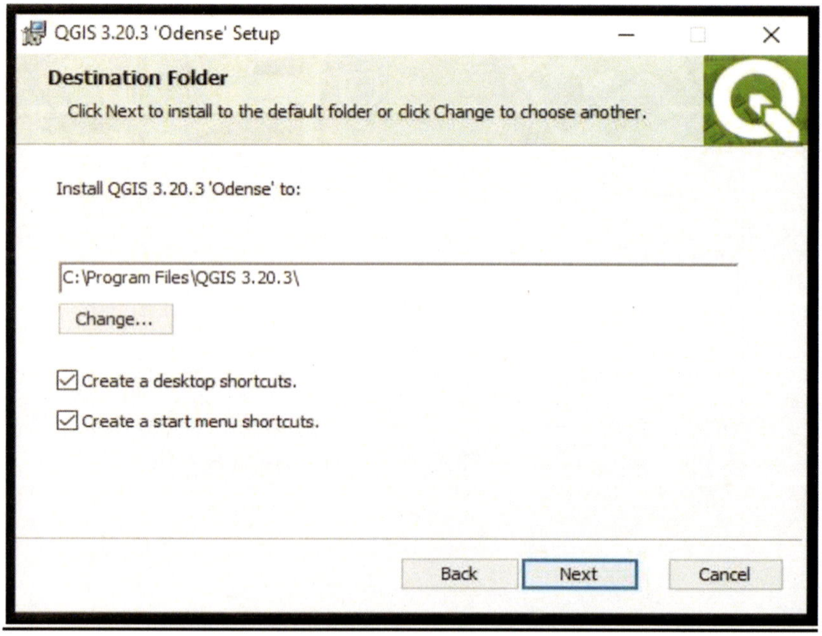
QGIS 3.20.3 'Odense' Setup
Destination Folder
Click Next to install to the default folder or click Change to choose another.
Install QGIS 3.20.3 'Odense' to:
C:\Program Files\QGIS 3.20.3\
Change...
Create a desktop shortcuts.
Create a start menu shortcuts.
Back
Next
Cancel

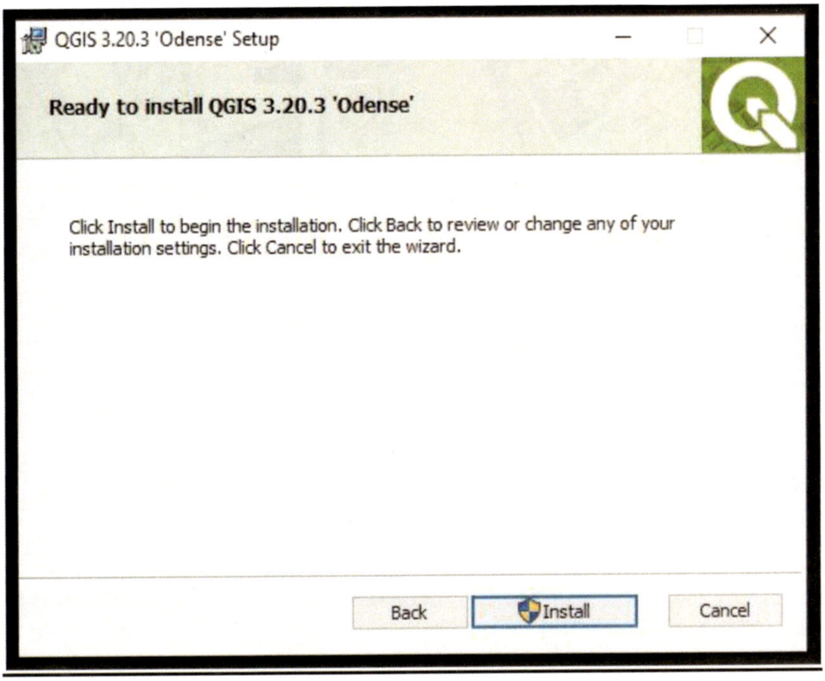
QGIS 3.20.3 'Odense' Setup
Ready to install QGIS 3.20.3 'Odense'
Click Install to begin the installation. Click Back to review or change any of your installation settings. Click Cancel to exit the wizard.
Back
Install
Cancel

QGIS 3.20.3 'Odense' Setup
Installing QGIS 3.20.3 'Odense'
Please wait while the Setup Wizard installs QGIS 3.20.3 'Odense'.
Status:
Back
Next
Cancel

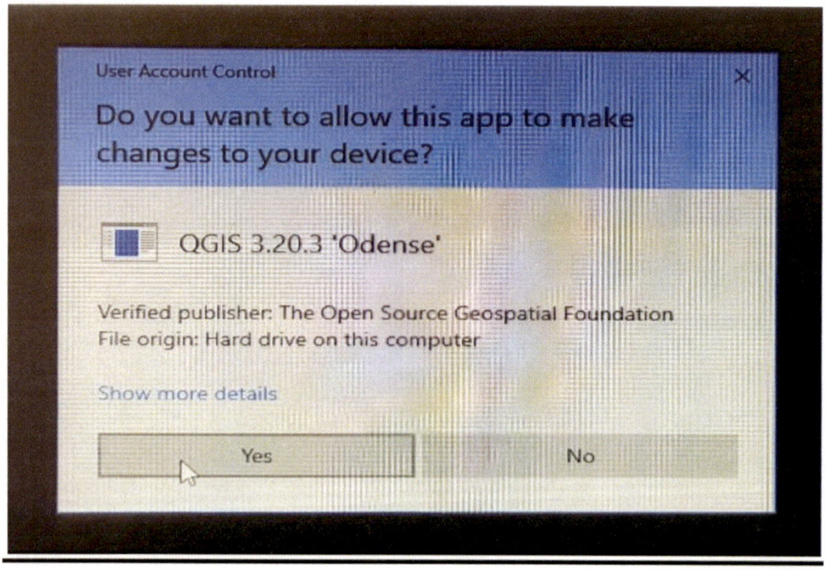
User Account Control
Do you want to allow this app to make changes to your device?
QGIS 3.20.3 'Odense'
Verified publisher: The Open Source Geospatial Foundation
File origin: Hard drive on this computer
Show more details
Yes
No

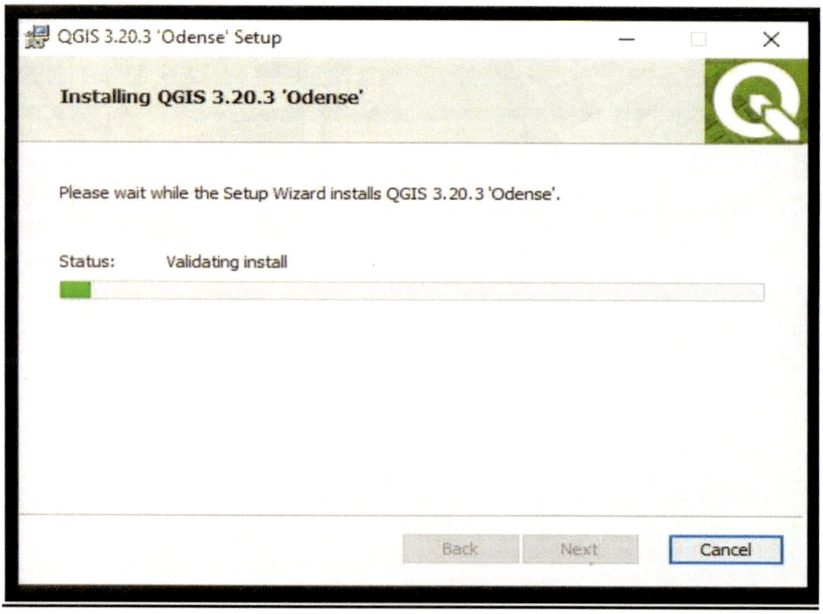
QGIS 3.20.3 'Odense' Setup
Installing QGIS 3.20.3 'Odense'
Please wait while the Setup Wizard installs QGIS 3.20.3 'Odense'.
Status: Validating install
Back
Next
Cancel

Visualization of Interface and Exploration of Tools

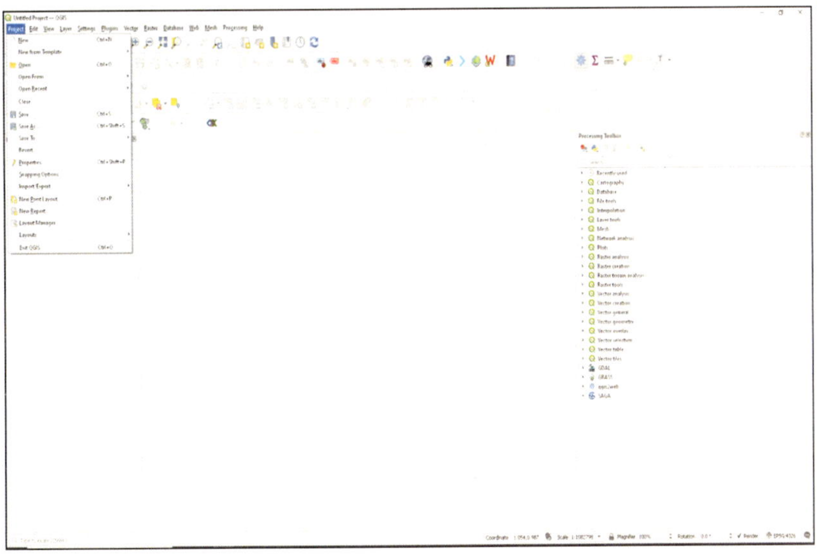

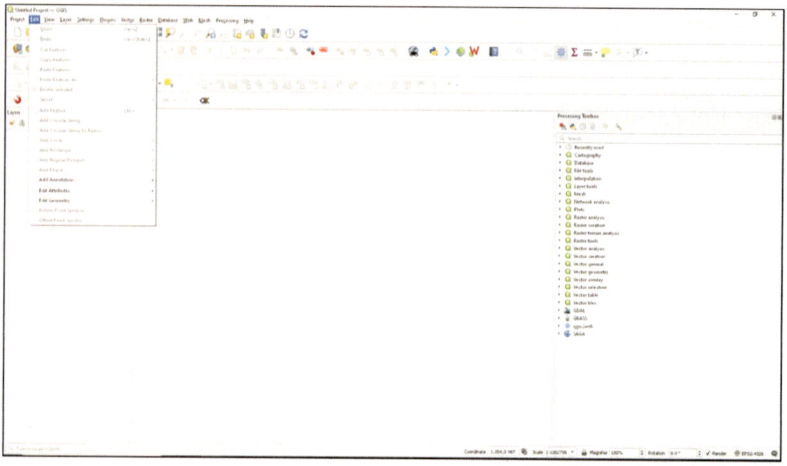

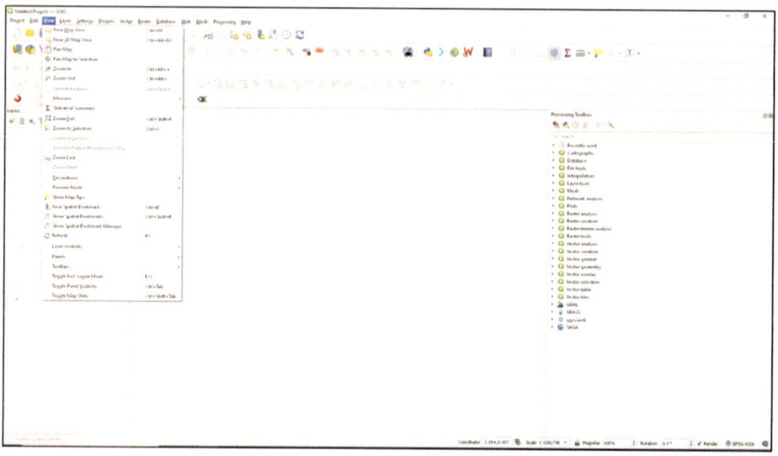

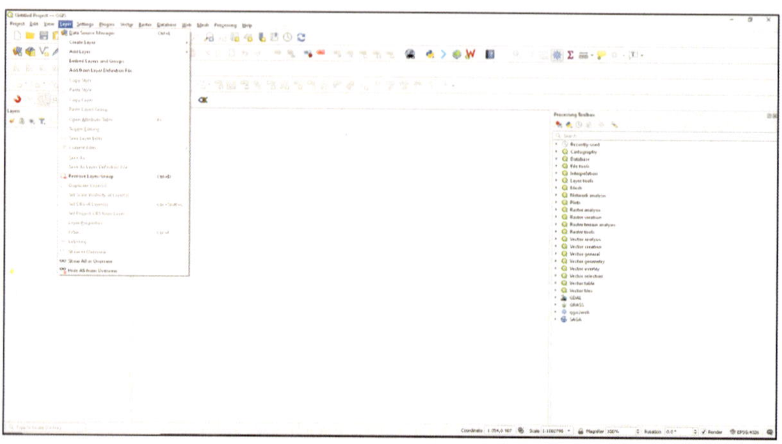

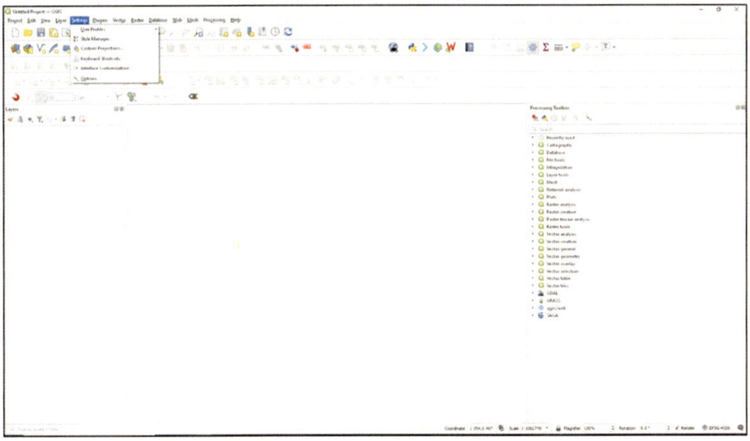

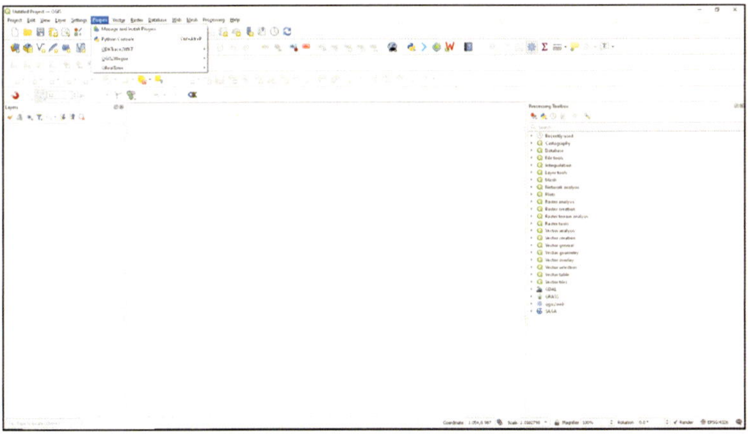

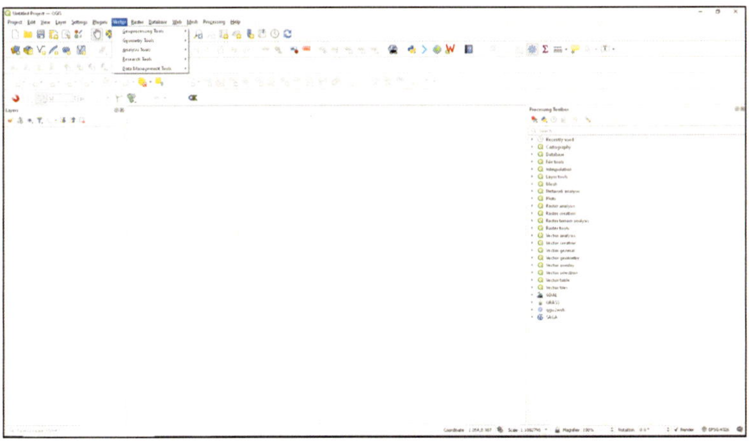

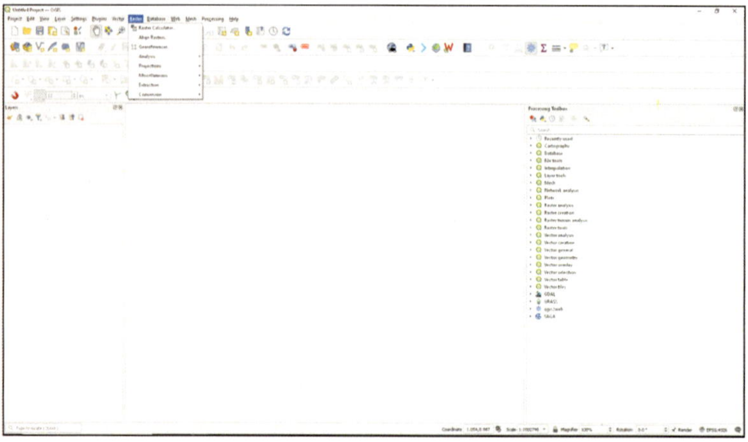

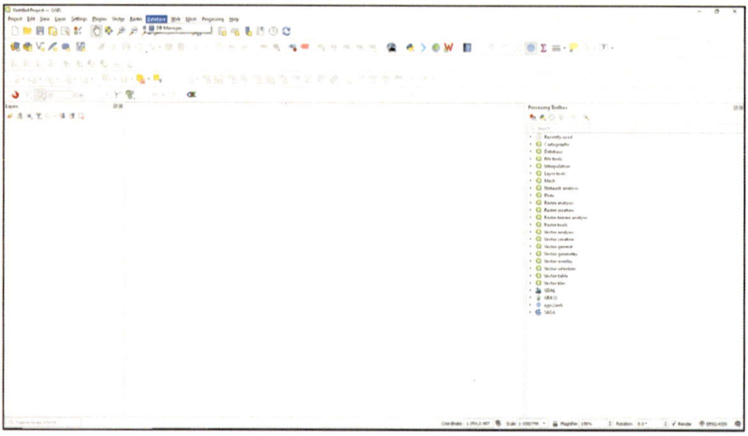

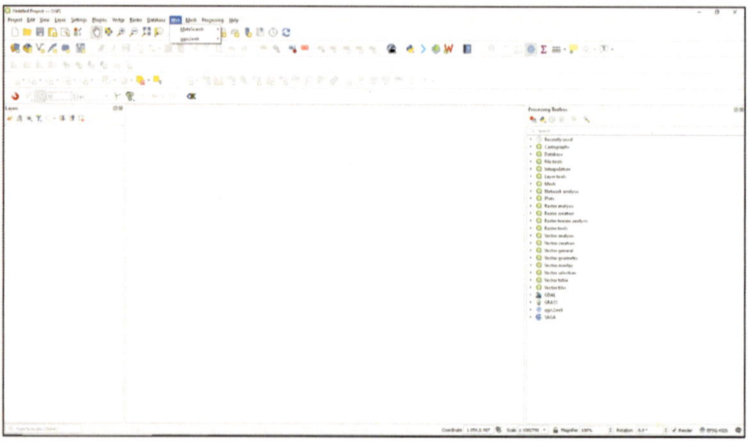

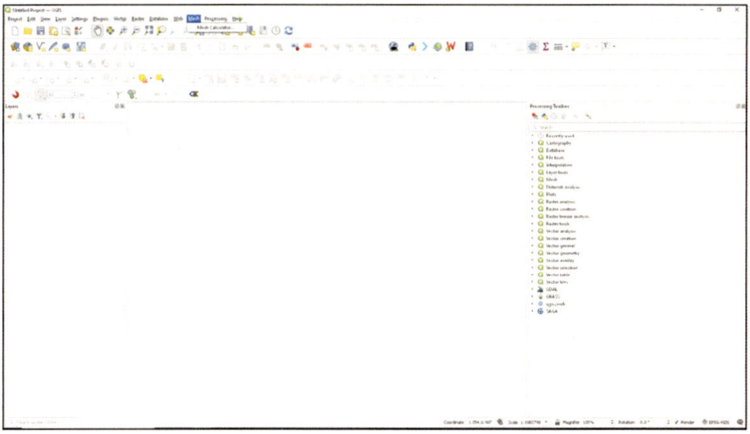

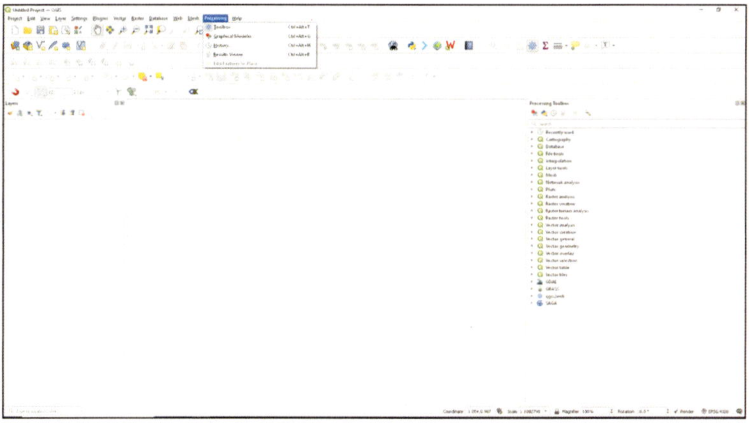

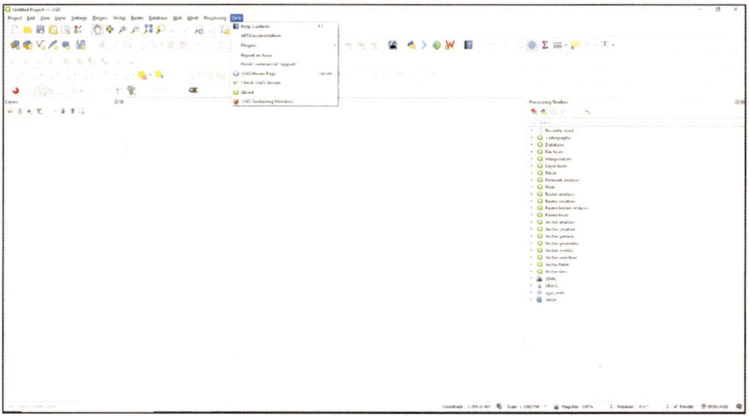

Importing SOI Map and Satellite Images

Theory

Both open source and commercial satellite imagery are available for download. Open source satellite imagery are mostly used from the LANDSAT and SENTINEL missions. Also, there are many satellite products like elevation data, rainfall data, vegetation data, etc. available for download. The following are links to download imagery.

Following are the links to websites from which you can download satellite imagery and products

- https://earthexplorer.usgs.gov/
- https://asf.alaska.edu/

Procedure

1. Click on *Add Raster Layer* tool in layer menu

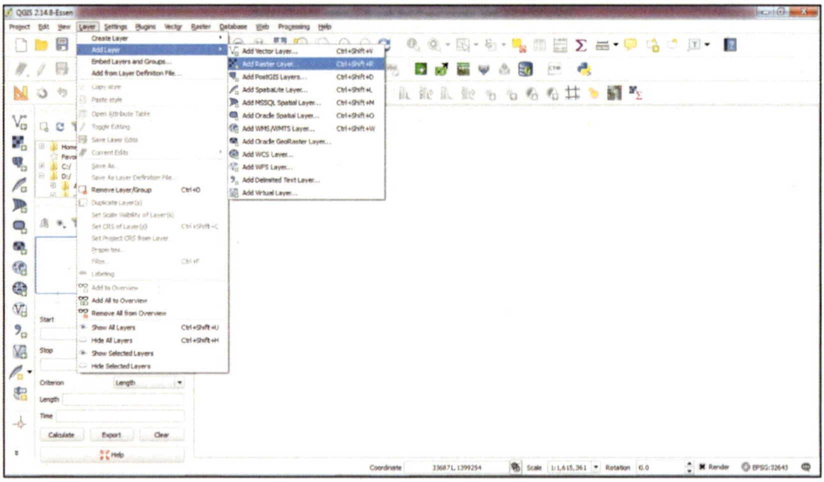

2. "Open a GDAL supported Raster Data Source" window that will pop out

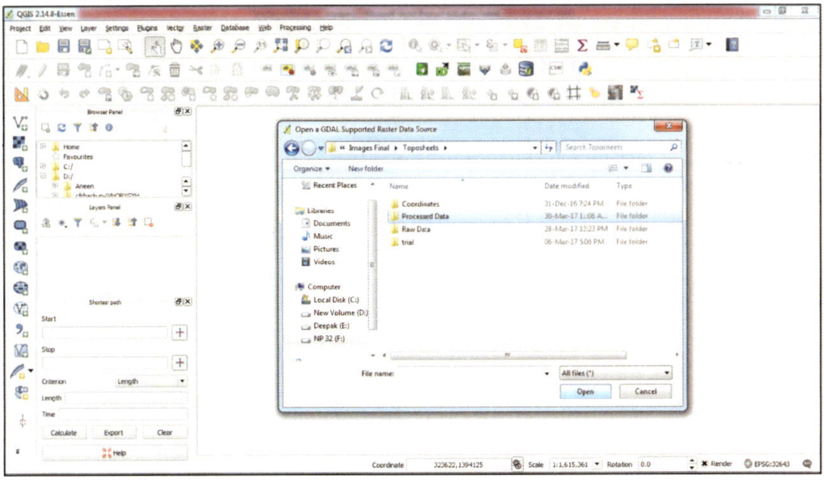

3. Browse and open the required Raster Data

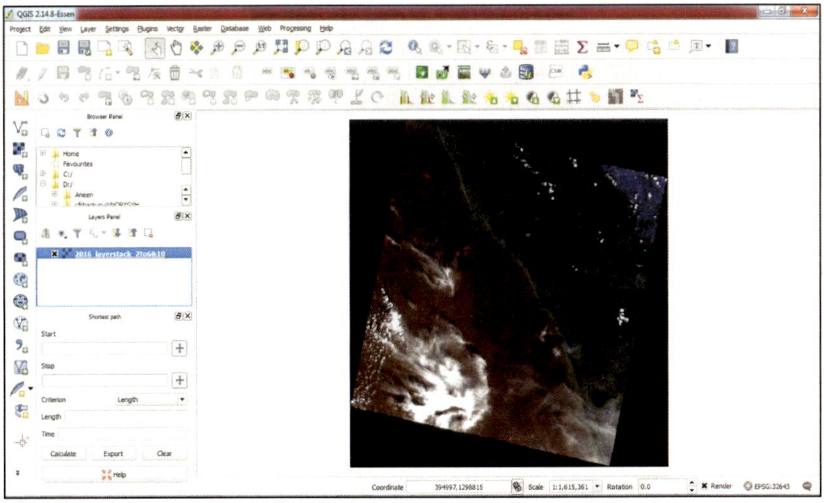

To change the band combination—Right click on the Raster image in layer panel and open its property. Band Combination can be changed in "Style" parameter.

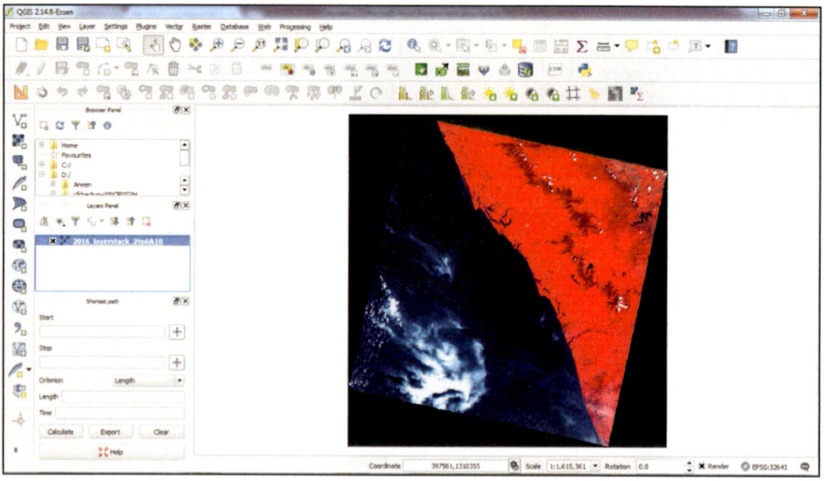

Band Combination changed to FCC. Similarly, SOI Toposheets can also be added.

4. To change the projection from Projected Coordinate System to Geographic Coordinate System or vice versa, follow the steps given below.
   a. Select the image
   b. Select "Project Properties" from "Project Menu"
   c. Enable the "on the fly" CRS transformation in the popped-up project properties table
   d. Select the required Projection

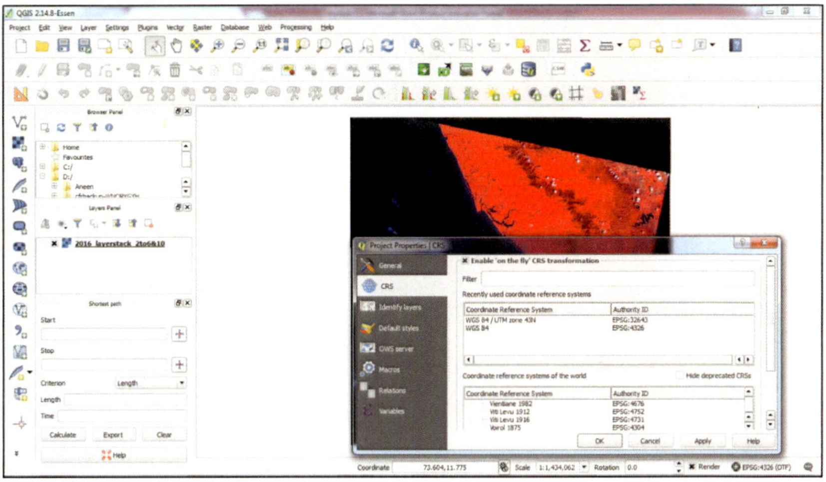

Geo-Referencing

Geo Referencing of Toposheet

Theory

It is often necessary to locate an image, or an aerial photograph in its correct spatial location and orientation for use as a backdrop in a GIS display or map. The name given to various methods used to achieve this is geo-referencing, which involves identifying of Ground Control Points (GCP) on both the image and in relation to a known datum. GIS software is then used to resample the image such that its spatial location and orientation matches with that of defined by the selected GCPs.

Procedure

1. Go to *Raster* → *Georeferencer*

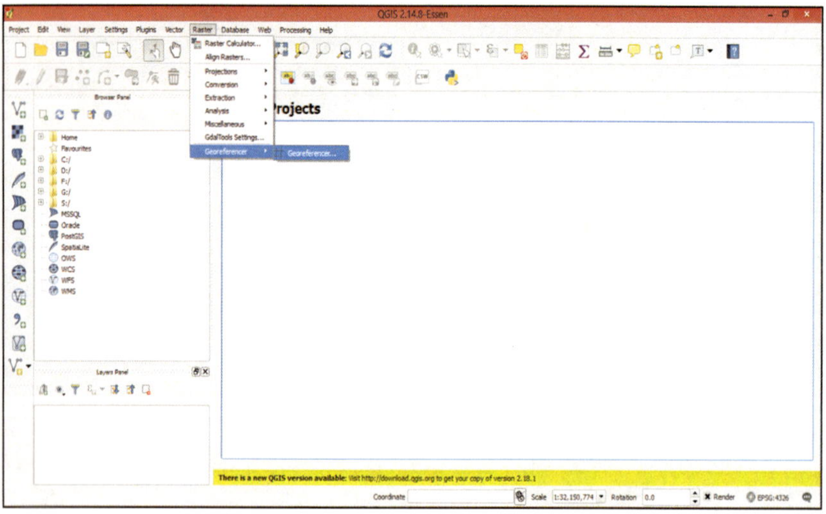

2. The georeferencer window will pop out

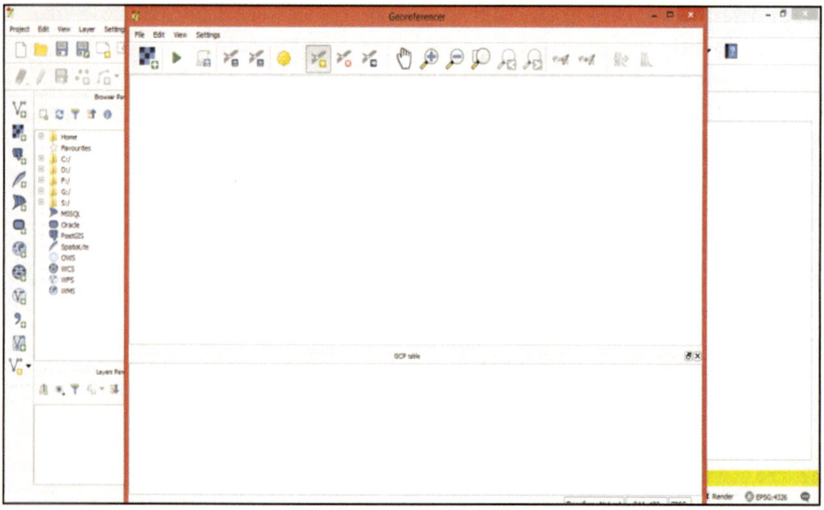

3. Add toposheet using the *Add raster layer* button 

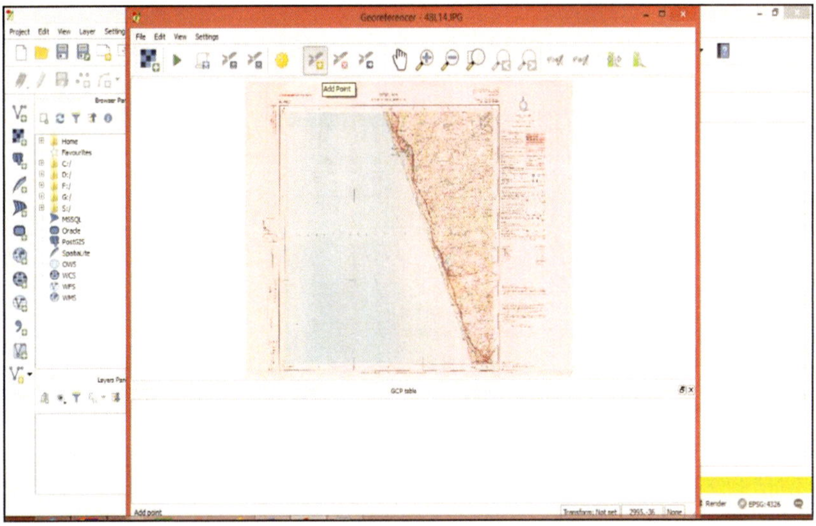

4. Click on the add point button . Click on the 4 coordinates and enter the corresponding values in the X and Y columns. Click on the Settings tab and then select Transformation settings

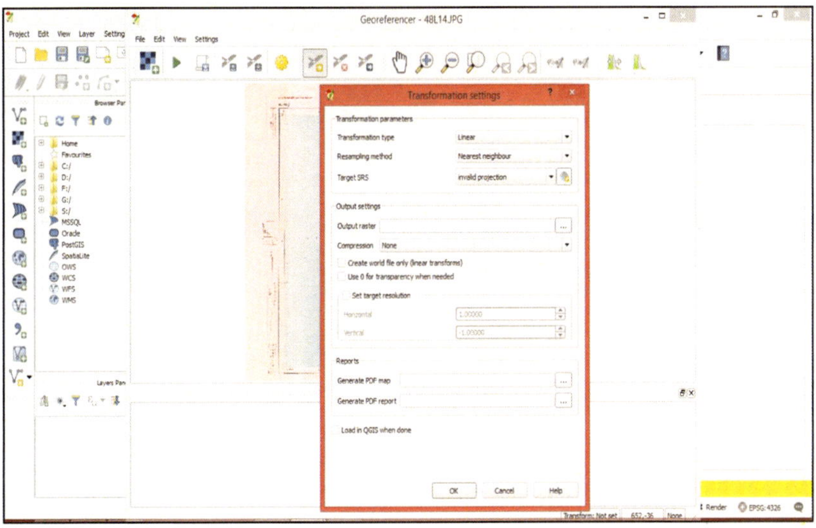

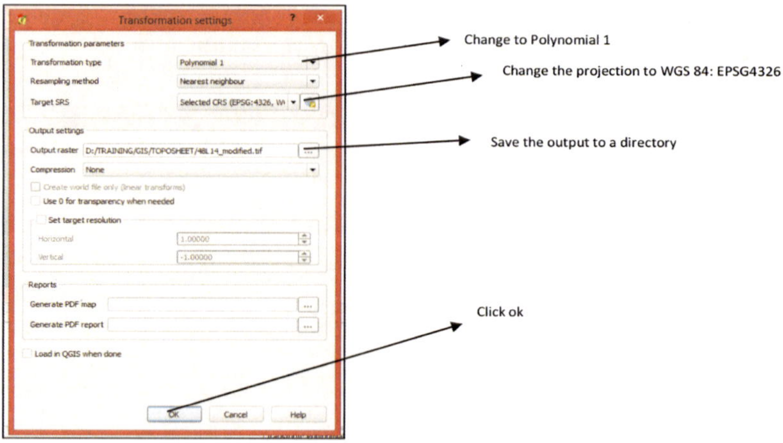

5. Click on the start georeferencing button . The image is now georeferenced

Image to Image Rectification

Procedure

1. Add Georeferencer from the main window, and then go to *Raster → Georeferencer*

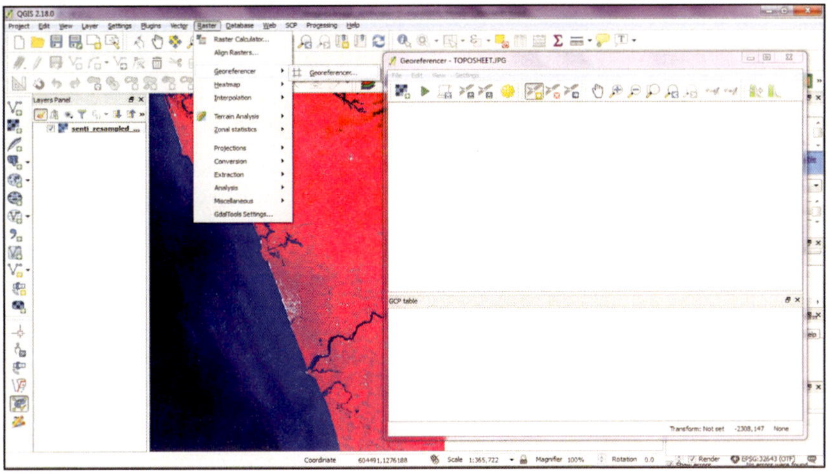

2. Add Raster in Georeferencer wizard using *Open Raster*  → click *Transformation setting*  → *OK*

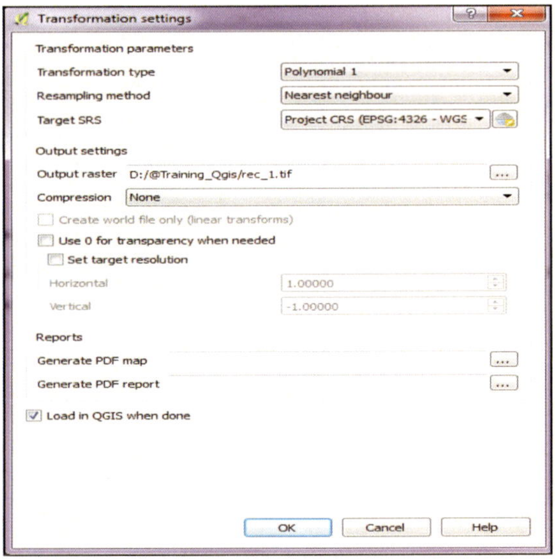

3. Click *Add point*  → Choose specific point using cursor, *Enter map coordinates* box will appear and then click *From map Canvas*

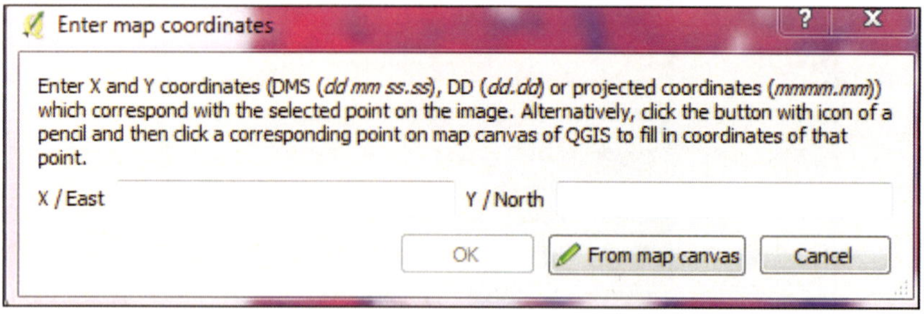

4. Now choose point from Georeferenced data, likewise collect as much as possible (Minimum 4 GCP points are required). Finally, click Start *Georeferencing*  →*Finished*

Sub-Setting of an Image

Theory

Often, from a country's vector file or shape file, we could need a specific section, like one state or district or grama panchayat or a defined boundary for study. In this section we tried to give step-wise procedure for removing a specific item/area from a layer/image and saving it as a new layer for analysis.

Procedure

1. Go to *Layer* → *Add vector layer*

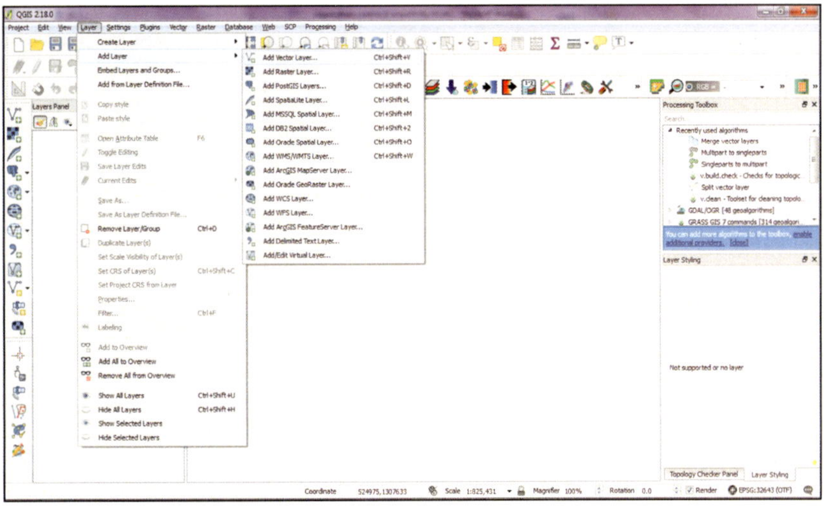

2. Add required shape file

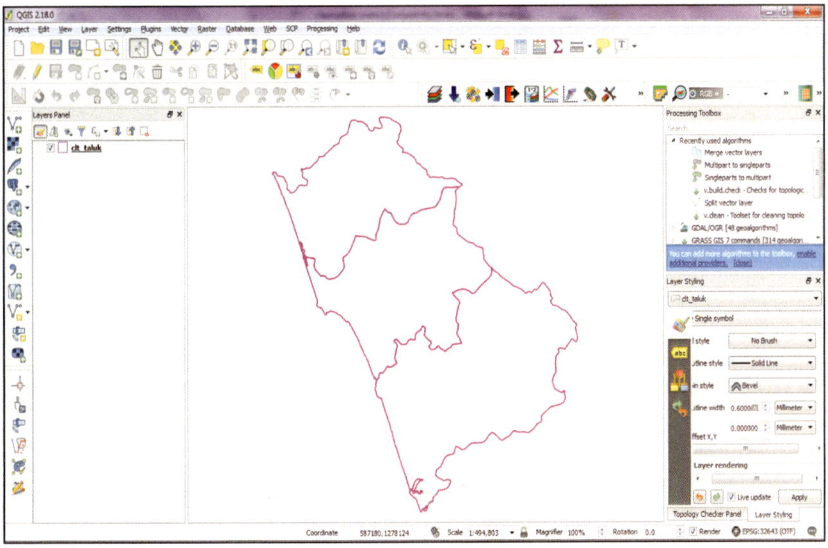

3. Enable  *Toggle editor* → Using  Select feature icon *Select AOI* → Click *Edit* → *Copy feature as* →  *Disable Toggle editor* → *Turn off layer*

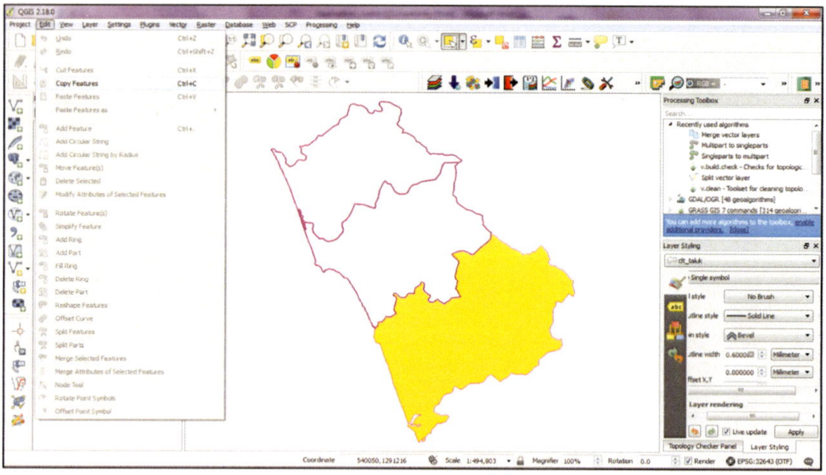

4. Go to *Layer* → *Create layer* → *New shapefile layer*

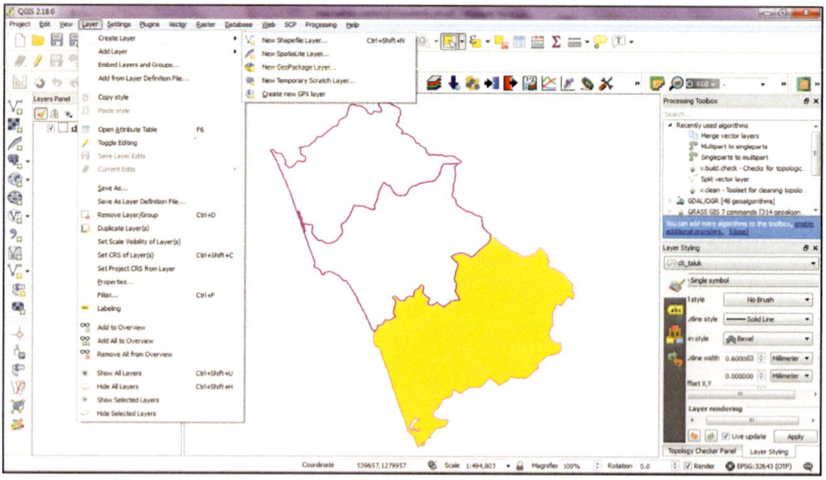

5. Turn on *AOI layer* → Enable *Toggle editor*

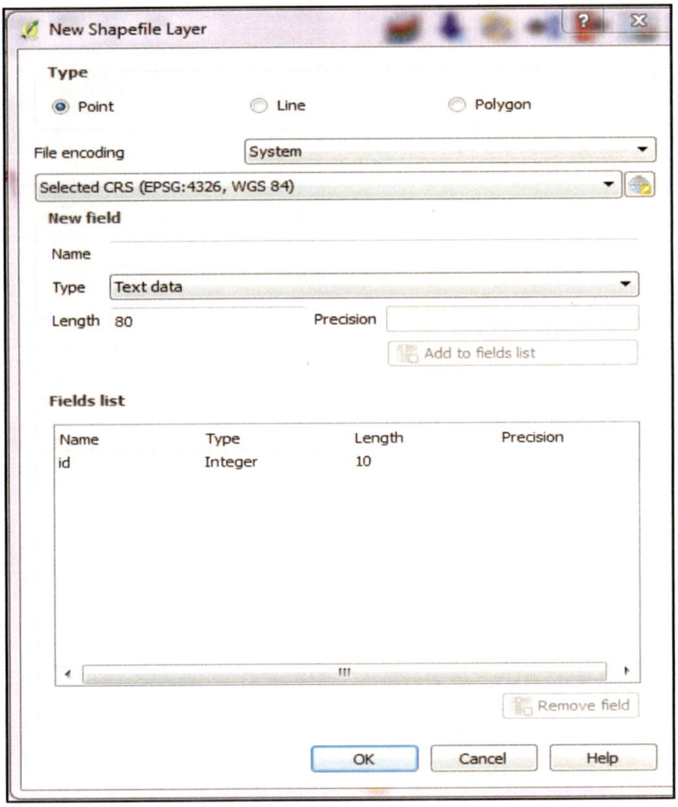

6. Go to *Edit* → *Paste layer as*

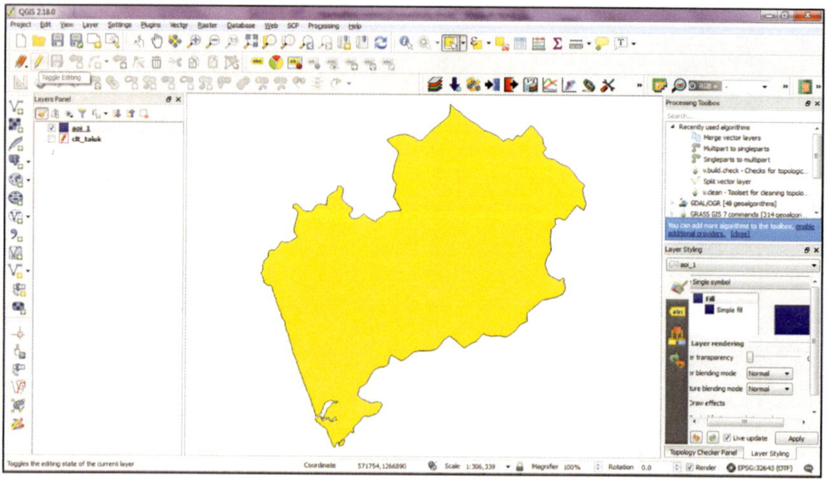

7. Go to Layer → Add Raster layer and overlay by vector layer

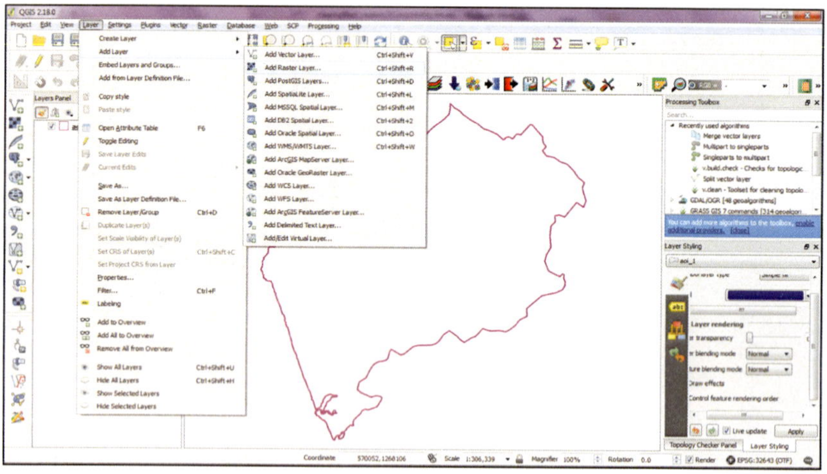

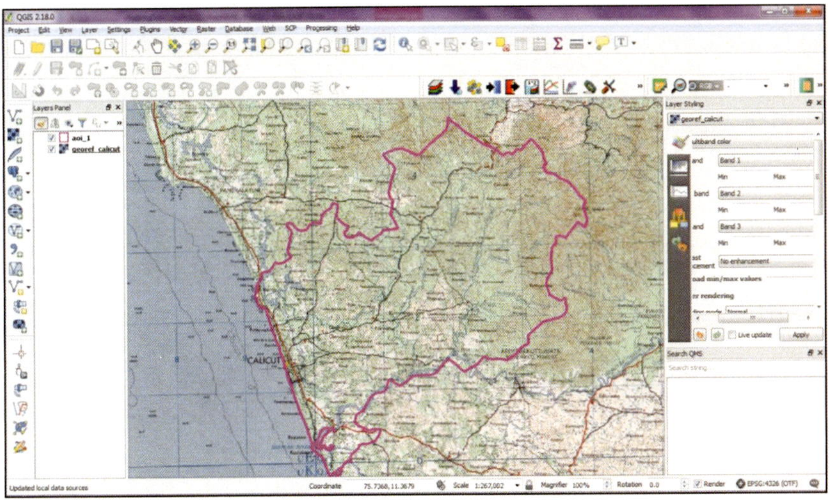

8. Go to *Raster* → *Extraction* → *Clipper* → *OK*

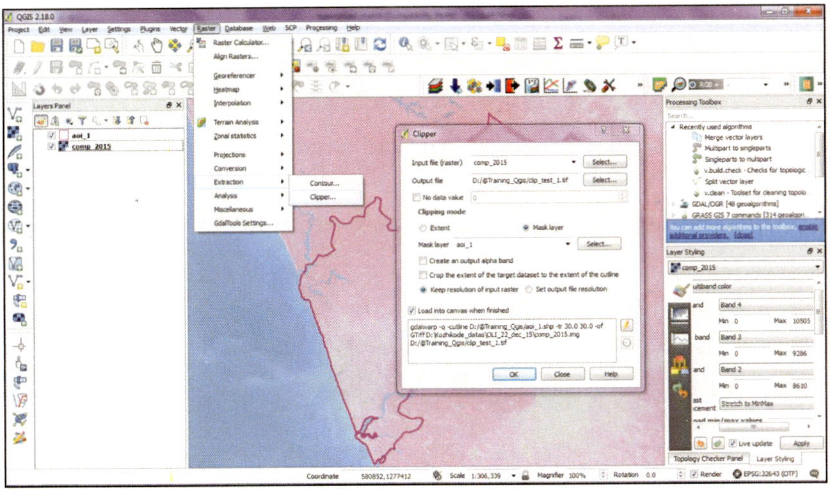

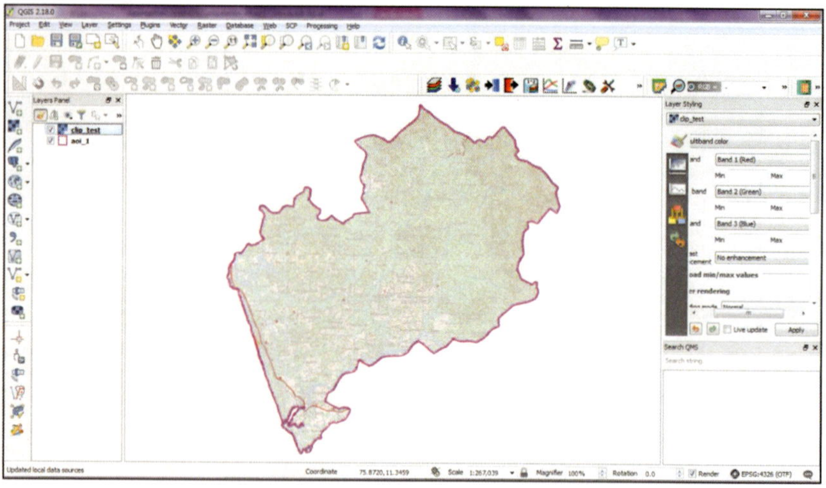

GIS Analysis

Creation of Point, Line and Polygon Features

Theory

A key method for data extraction and storage in GIS development is through on screen digitization. It is used to record coordinates/ locations in the form of points, lines, or polygons.

Procedure

1. Go to *Layer → Add layer → Add Raster layer (Raster layer may be your toposheet, satellite image located in the drive of your computer)*

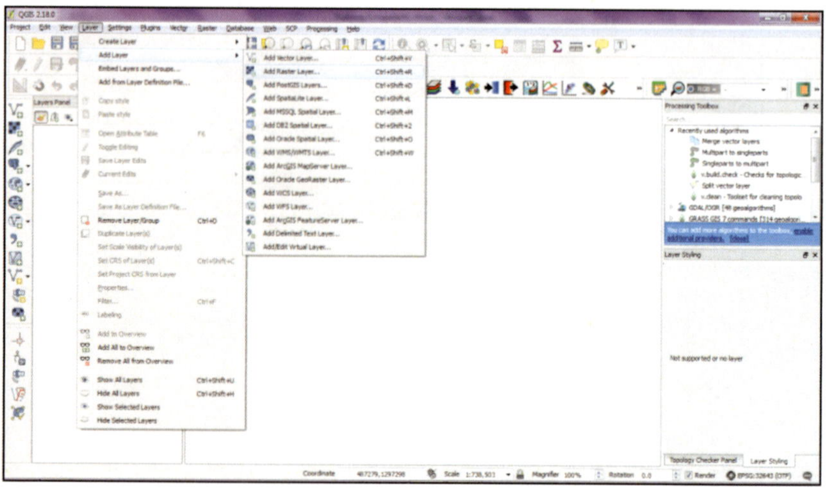

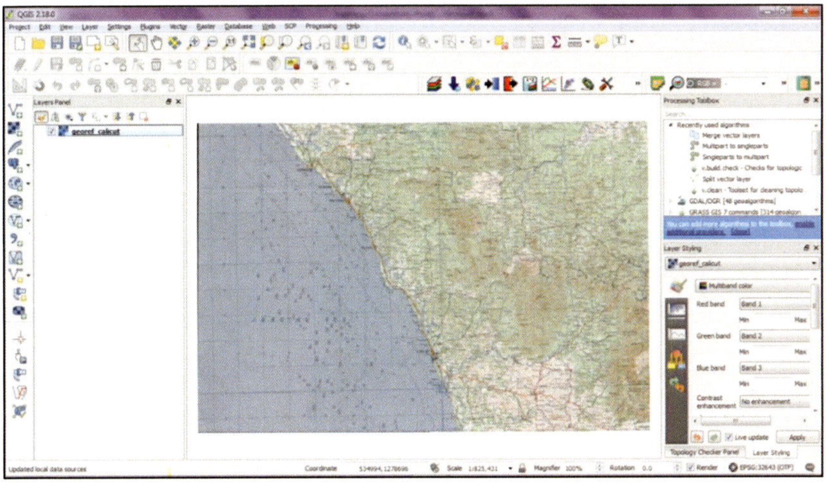

2. Go to *Layer* → *Create layer* → *New Shape file layer*

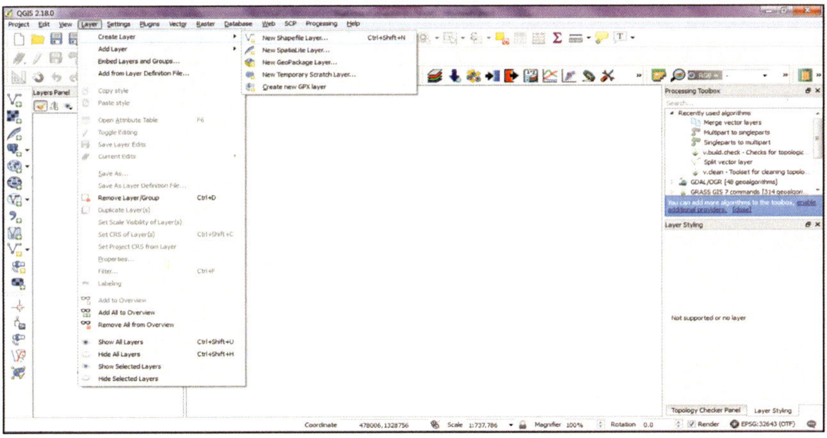

3. Creating Point, Line, and Polygon Features

   i. Creating Point Feature

   After creating the shape file, go *to Layer → Select Test_Point → Enable Toggle editor*

   ii. Creating Line Feature

   After creating the shape file go to *Layer → Select "Test_line" → Enable Toggle editor*

iii. Creating Polygon Feature

After creating the shape file go to *Layer → Select "Test poly" → Enable Toggle editor*

Topology Errors and their Rectification

Theory

The relationships between the points, lines, and polygons that constitute a geographic region's features are referred to as topology. Using the plugin **TopologyChecker**, several topology criteria/ rules can be used to examine your vector files and determine their topology.

Procedure

1. Install "Topology checker plugin"

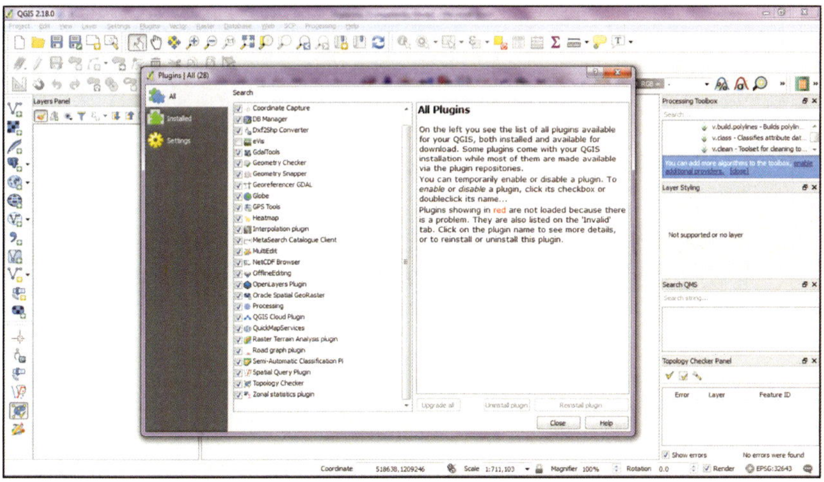

2. Add Vector Layer, and then go to Vector → Select Topology checker

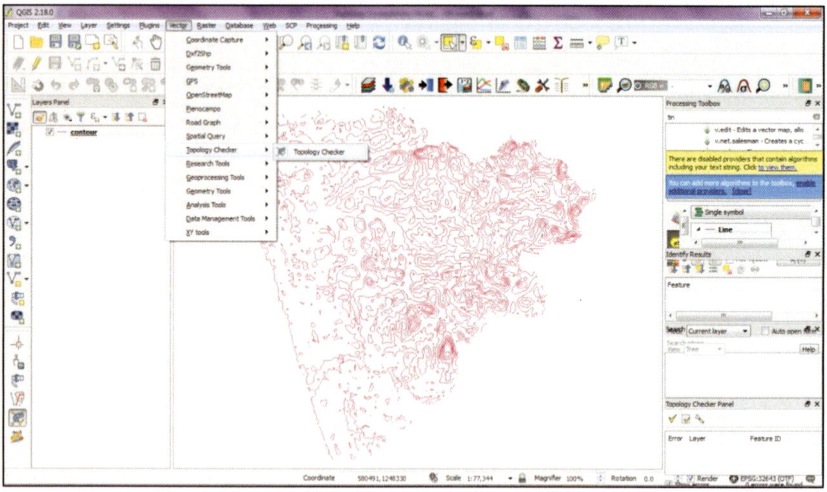

3. Then go to "Topology Checker panel" click Configure option → Fill settings Add Rule → give "OK"

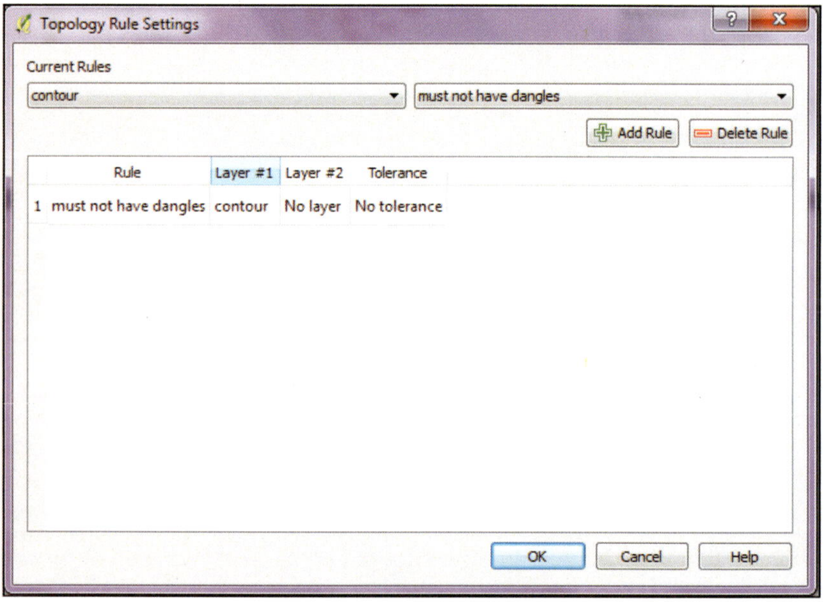

4. After that click Validate all option → then the error will be highlighted in RED color

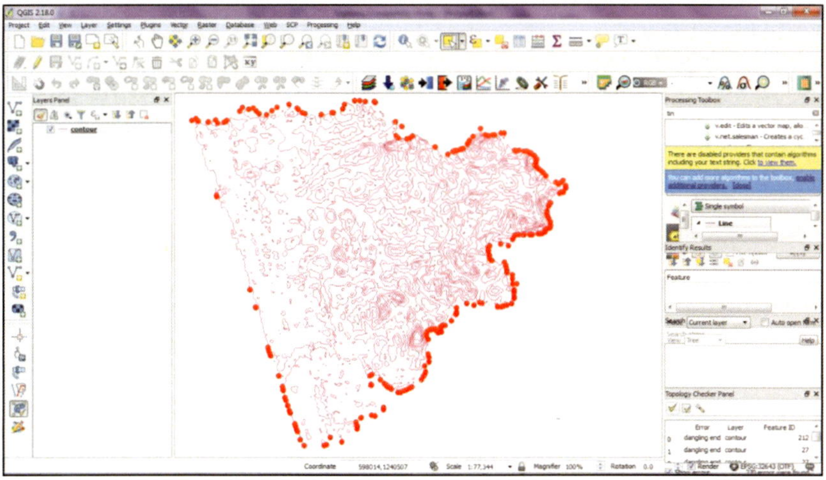

5. Then click v.clean - Toolset for cleaning topology of vector map. option → fill settings → Run

   After that click “Validate Extent ” option → Finished

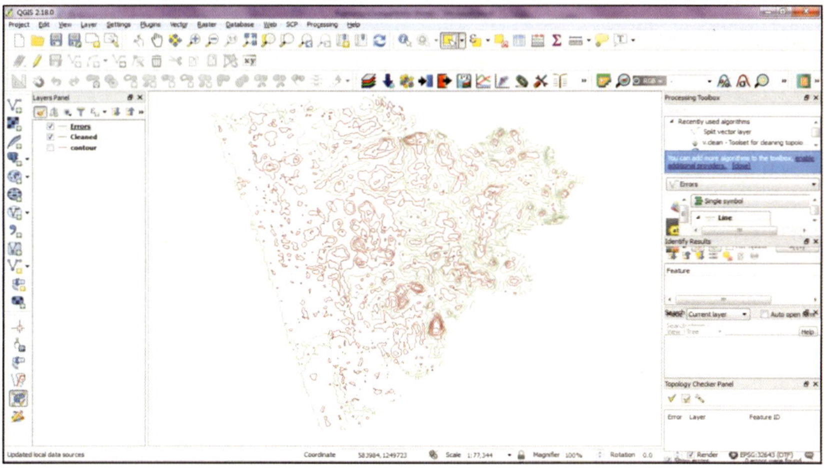

Assigning and Linking of Attributes

Theory

The attribute table displays information on the features of a selected layer. Each row in the table represents a feature (with geometry or not), and each column contains a particular piece of information about the feature. Features in the table can be searched, selected, moved, or even edited.

Procedure

Open Attribute Table

1. First open "vector file" → Right click on the particular file → Select "Open Attribute Table"

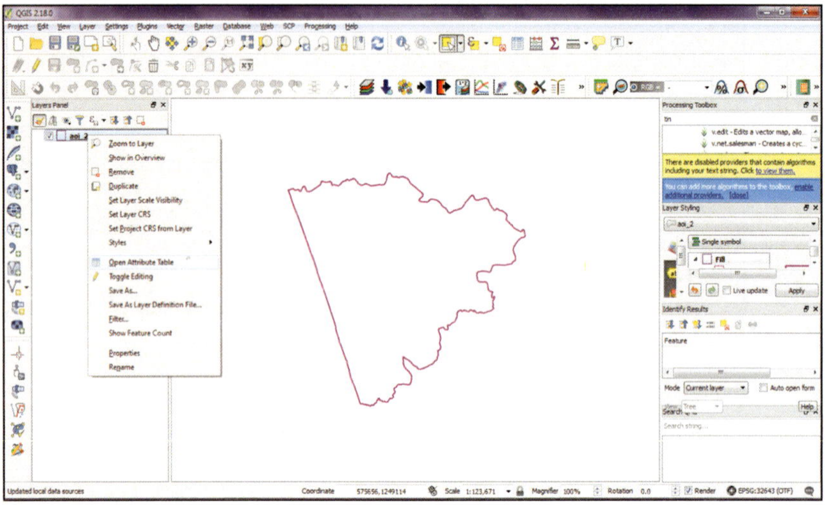

Edit Attribute Table

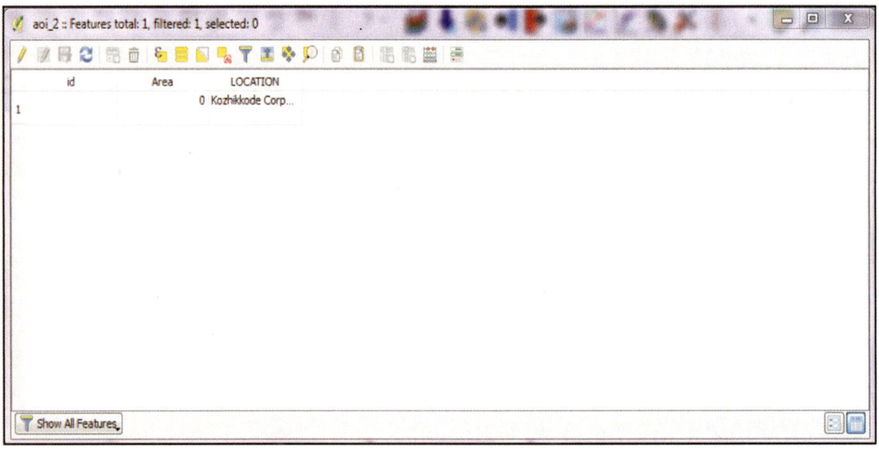

Linking Of Attribute Table

1. Open necessary layers, then select target layer → Right click "Properties" → Select "Join tab"

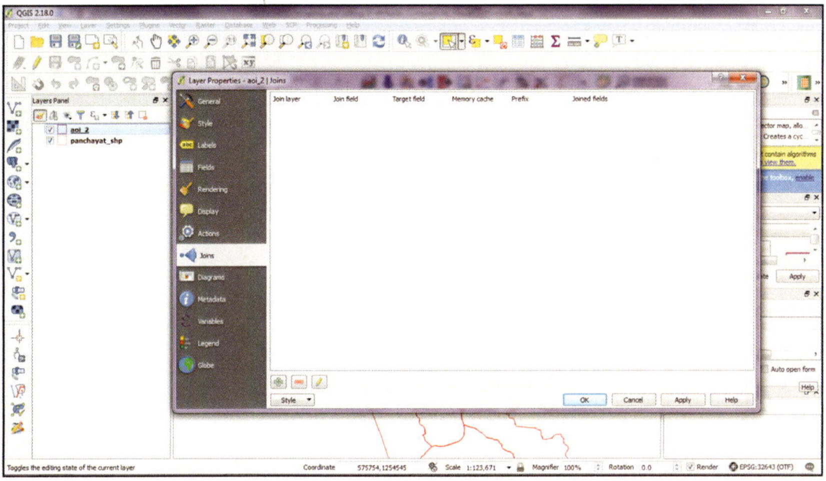

2. Using the "Add vector Join" option to link two vector attributes

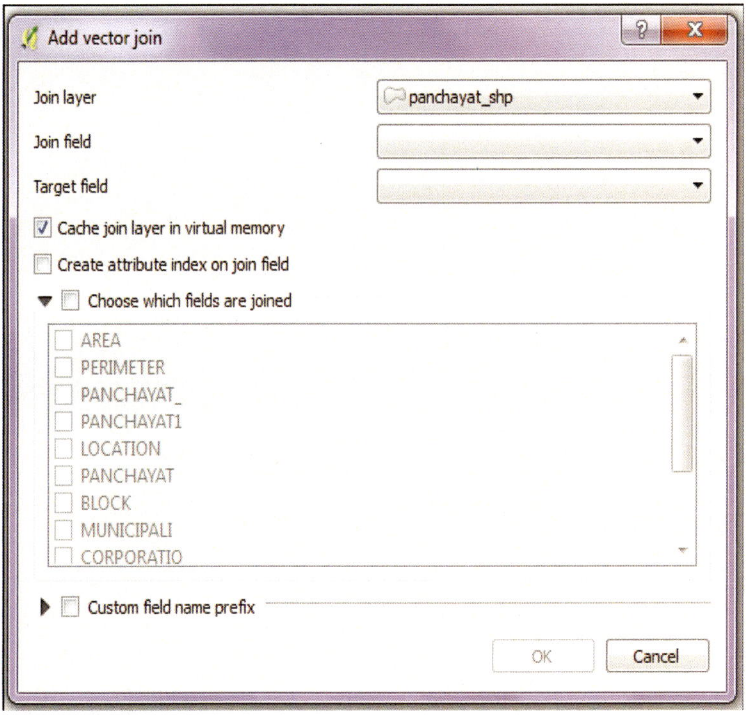

Stream Delineation Using Digitizing Technique

Procedure

1. To add raster file, go to Layer → Add Raster Layer

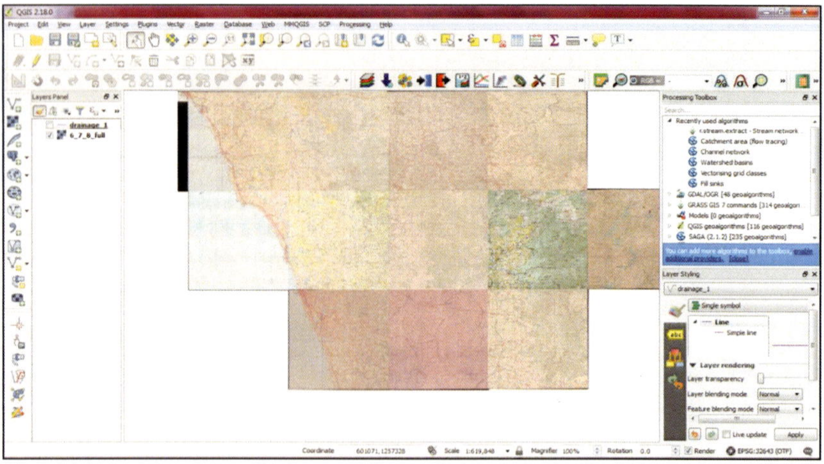

2. Create New Shape file (Line), Layer tab → Create New Layer → New Shapefile Layer. Fill in the fields and click OK

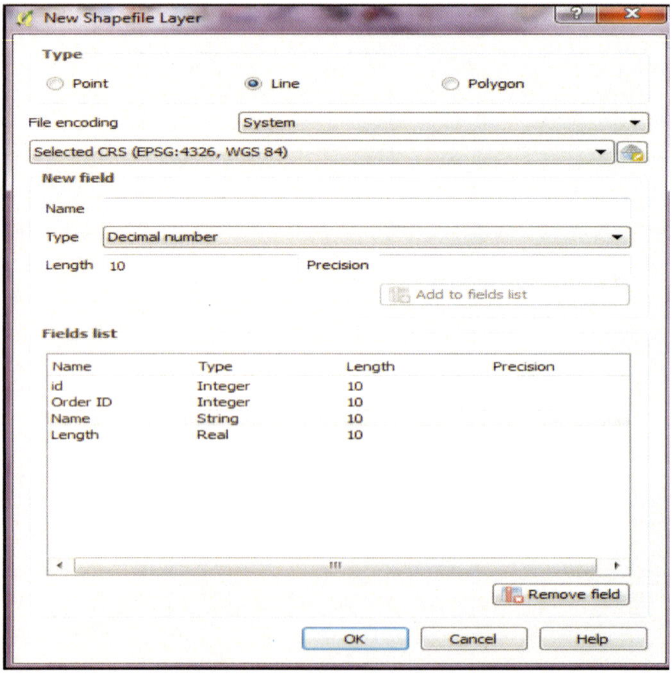

3. During digitization, enter Order ID

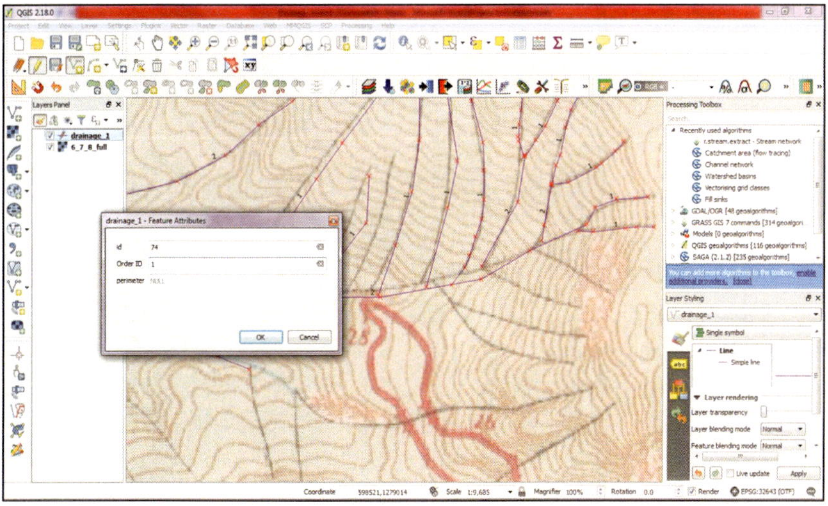

4. After digitization, go to "Open Attribute Table" → Open "Field calculator" → create "length" column → OK

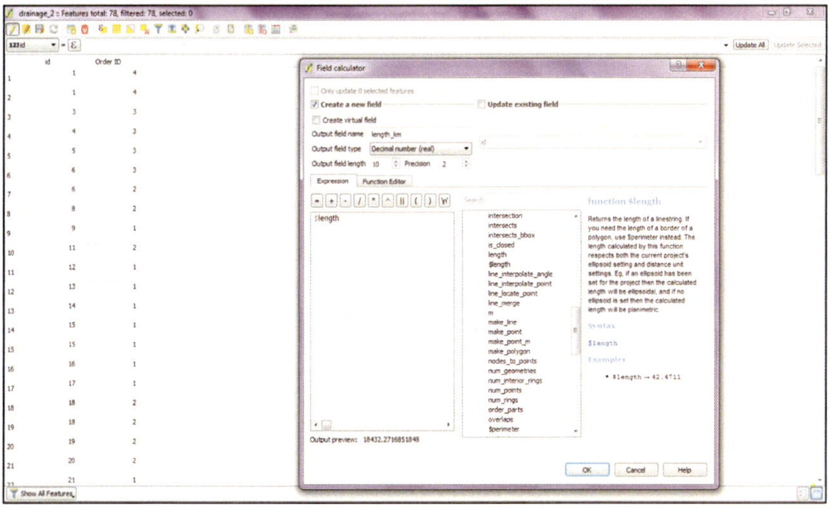

5. For length statistics, go to vector tab → Analysis Tools → Select "Basic Statistics for numeric fields" → Run

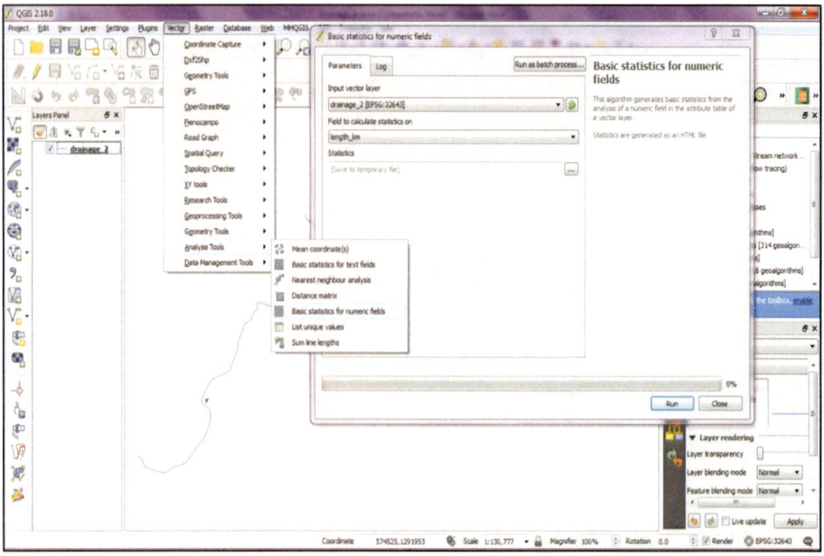
Basic statistics for numeric fields
Parameters
Log
Run as batch process...
Input vector layer
drainage_2 [EPSG:32643]
Field to calculate statistics on
length_km
Statistics
Basic statistics for numeric fields
This algorithm generates basic statistics from the analysis of a numeric field in the attribute table of a vector layer.
Statistics are generated as an HTML file
Mean coordinate(s)
Basic statistics for text fields
Nearest neighbour analysis
Distance matrix
Basic statistics for numeric fields
List unique values
Sum line lengths
Run
Close

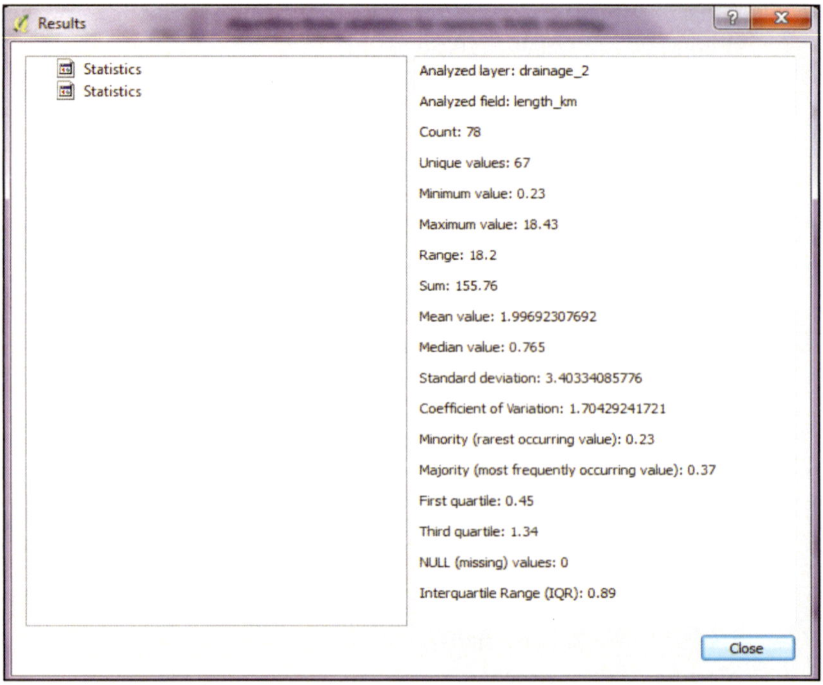
Results
Statistics
Statistics
Analyzed layer: drainage_2
Analyzed field: length_km
Count: 78
Unique values: 67
Minimum value: 0.23
Maximum value: 18.43
Range: 18.2
Sum: 155.76
Mean value: 1.99692307692
Median value: 0.765
Standard deviation: 3.40334085776
Coefficient of Variation: 1.70429241721
Minority (rarest occurring value): 0.23
Majority (most frequently occurring value): 0.37
First quartile: 0.45
Third quartile: 1.34
NULL (missing) values: 0
Interquartile Range (IQR): 0.89
Close

6. Final output

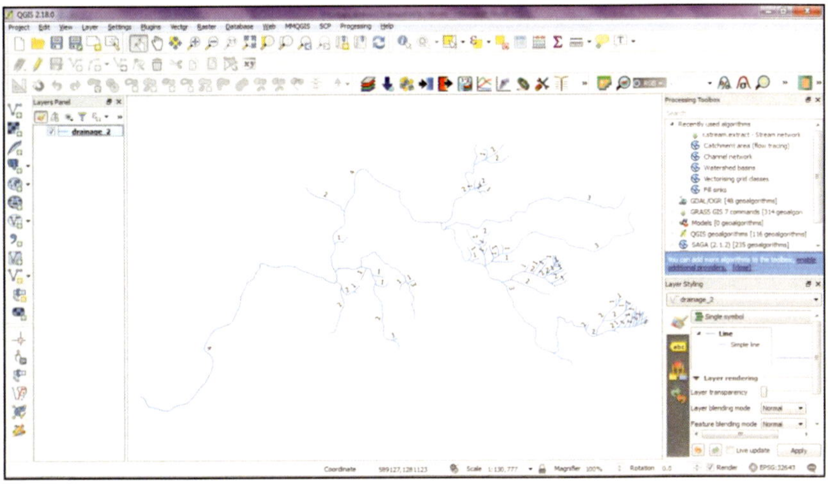

Union Analysis

Theory

Union is a mathematical function that is defined on two or more feature classes which outputs the collection of features in input feature classes. For example, if you have a point layer containing paddy field locations and another point layer containing locations of wells, then the operation that combines features in both the layers (paddy locations and well locations) is called union.

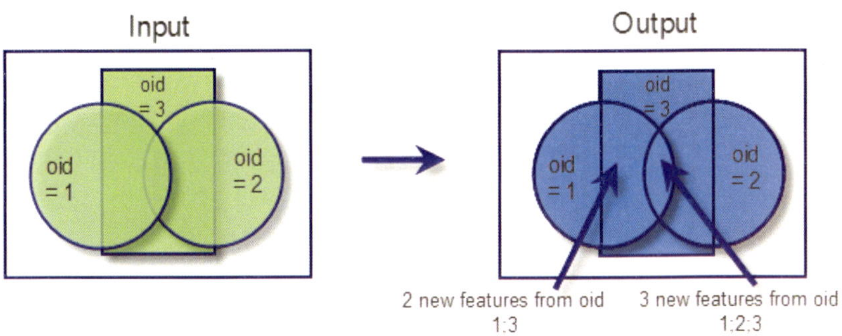

Source: How Union works—Help | ArcGIS for Desktop

Procedure

1. Go to Layer → Create Layer → New Shapefile Layer

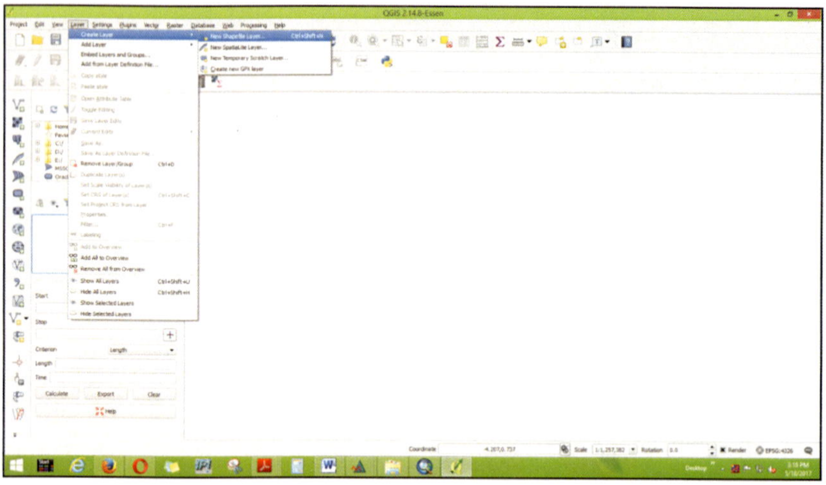

2. "New Shapefile Layer" window pops up. Create a required Shapefile with the necessary field list.

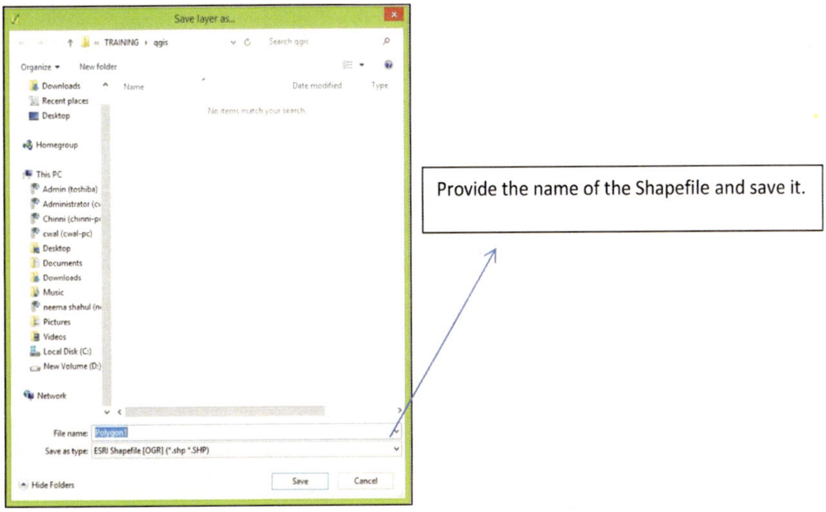

You can do union operation either by creating new Shapefiles or by adding an already existing shapefiles.

3. Similarly create or open the second set of Shapefile (as shown in image)

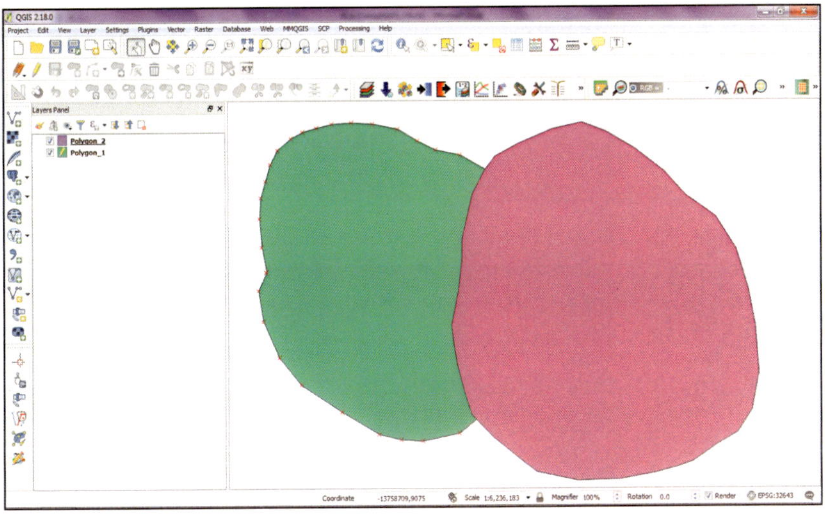

4. Go to Vector → Geoprocessing Tool → Union

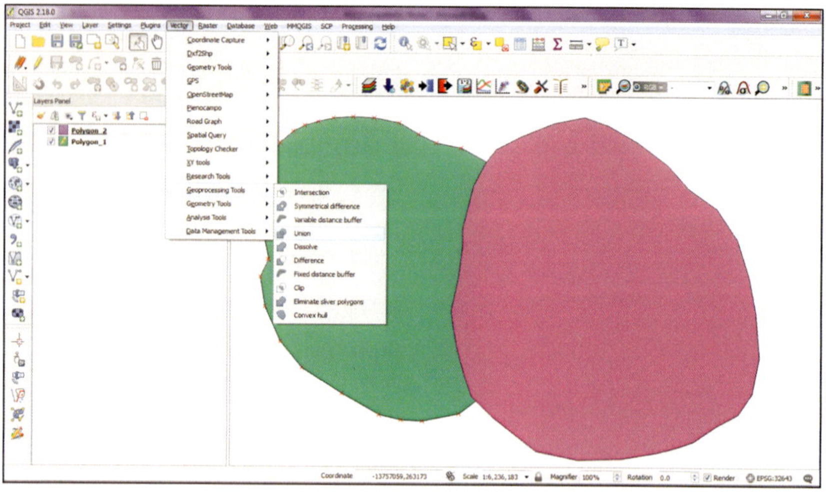

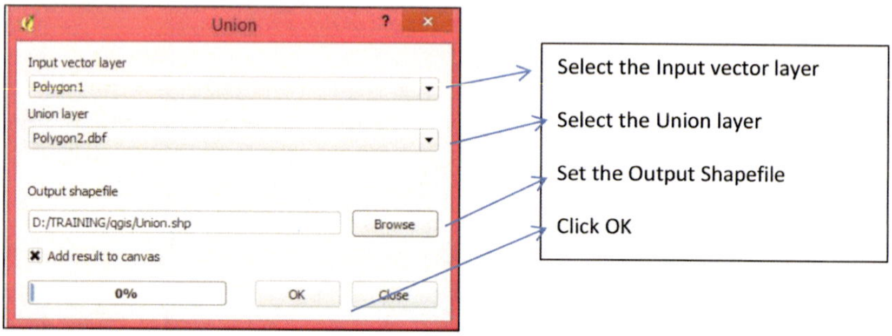

5. Union operation is done

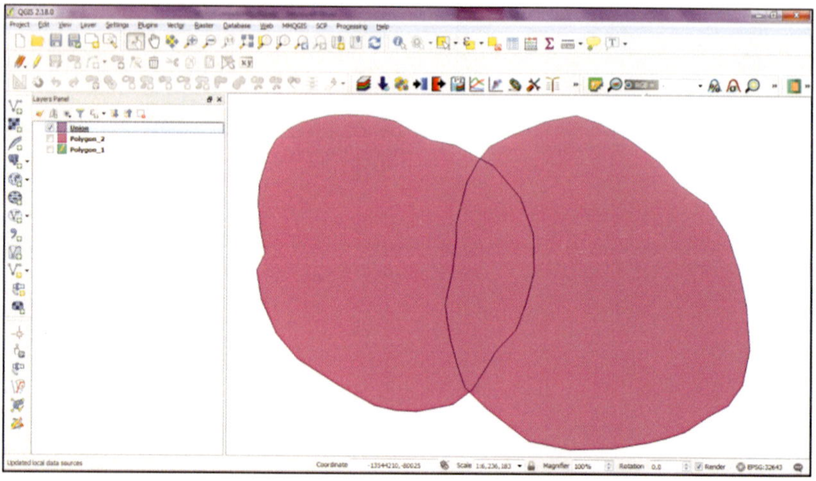

Intersect

Theory

Intersection is the tool that generates an output layer having the common objects in the input layers. For example, you have a polygon mapped with paddy field (example, district level) and you have another polygon layer of a particular Taluk in that district, then the intersection of these two layers will give paddy field locations in that particular Taluk.

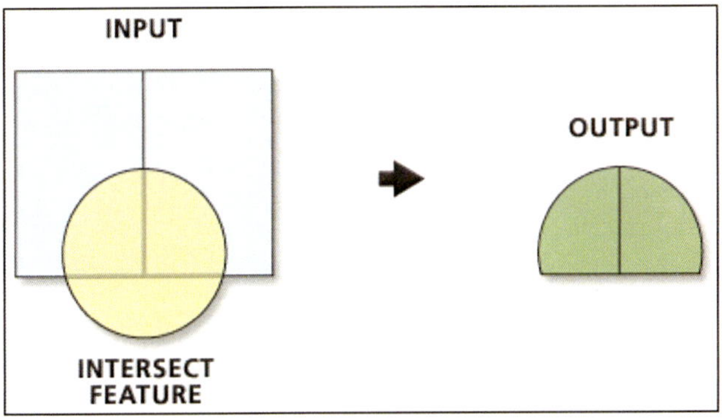

Intersection *source*: Intersect (Analysis)—ArcGIS Pro | Documentation

Procedure

1. Follow the same steps of Union operation and then select the "Intersection" tool in "Geoprocessing Tool" under "Vector" Layer

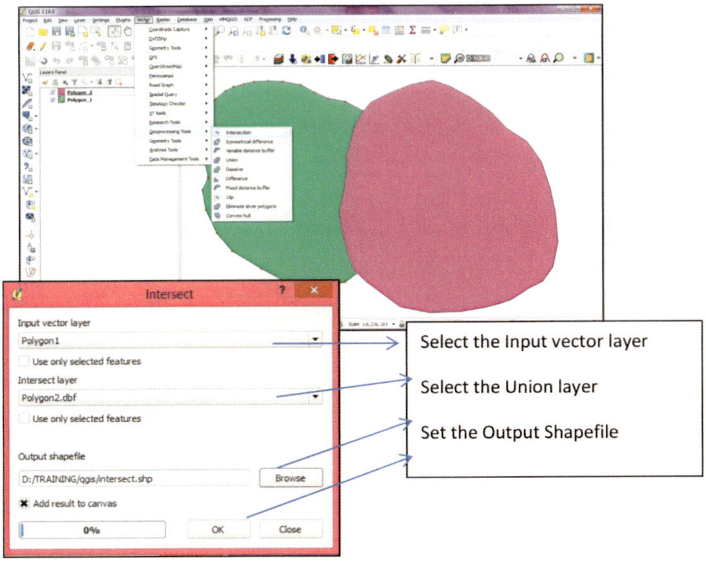

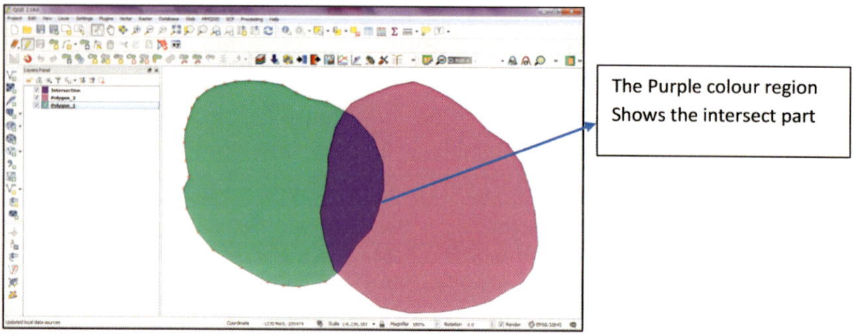

2. Intersection operation is done

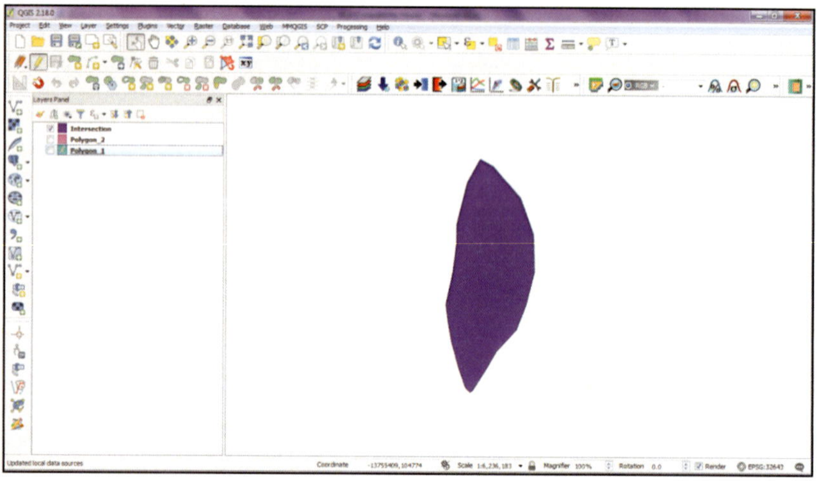

Buffer

Theory

Buffer analysis is used for identifying areas surrounding geographic features or specific areas of interest within desired distance. The process involves generating a buffer around existing geographic features and then identifying or selecting features based on whether they fall inside or outside the boundary of the buffer. The figure below shows the computation of the buffer for a linear feature.

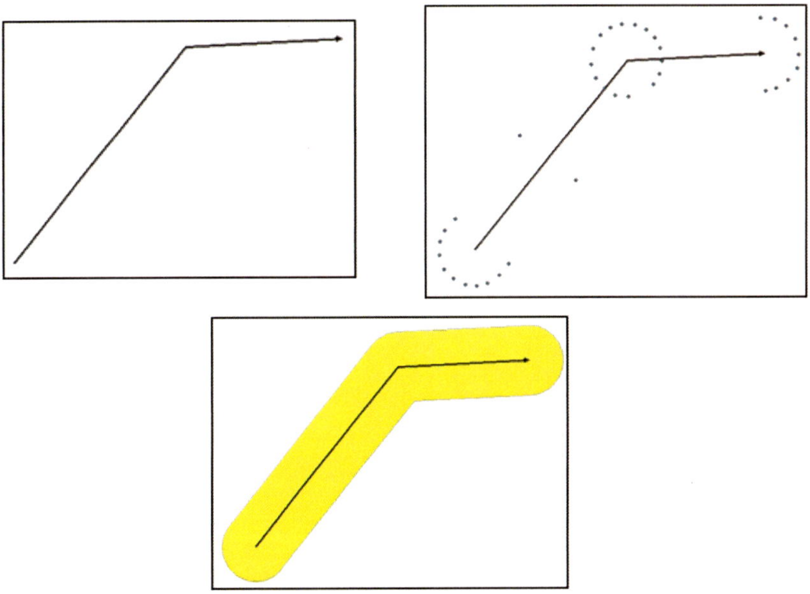

Buffer Analysis *Source*: How Buffer (Analysis) works—ArcGIS Pro | Documentation

Procedure

1. Click on the add vector layer button and browse to the location where the shapefile is located
2. Click on the vector tab, move to geoprocessing tool, and click on the buffer tab

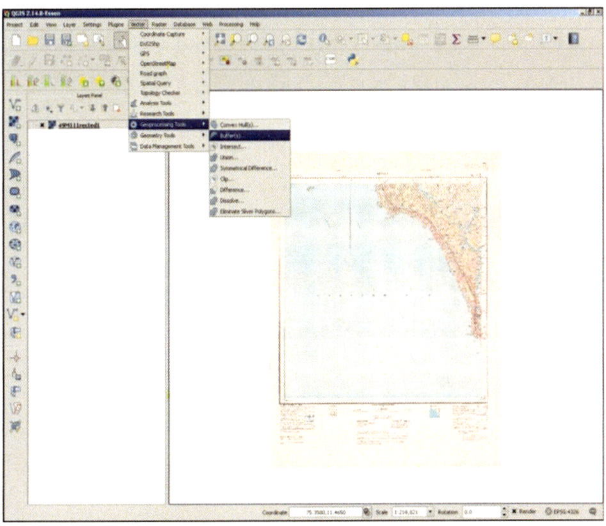

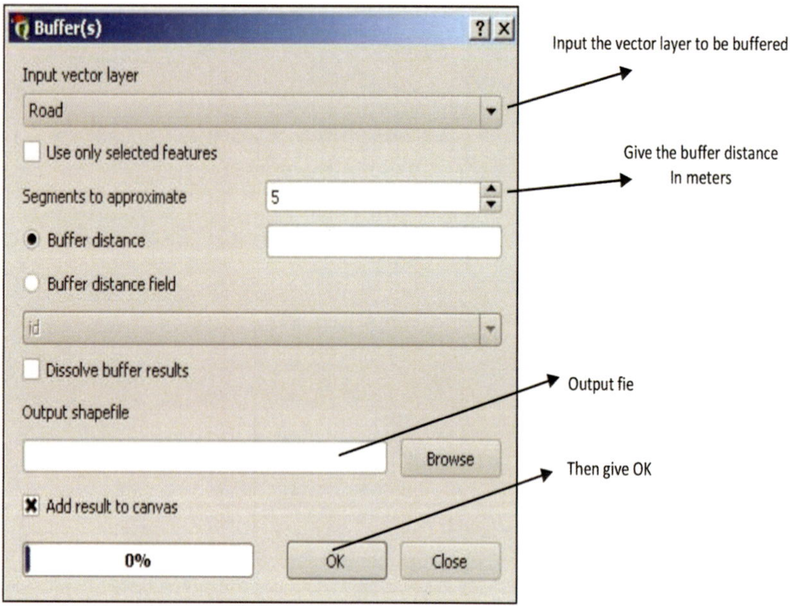

3. You can see the buffered region along your shapefile

CLIP

Theory

Clip is a tool that is used to extract a particular area of interest. Here, we have a land use shape file of a district (say X) and we require only the land use details of a particular Grama Panchayat in that district (say X-*a,* in white color). For this purpose, we use the Clip Tool and the procedure is shown below.

Procedure

1. From Vector Tab, select Geoprocessing Tools – Clip

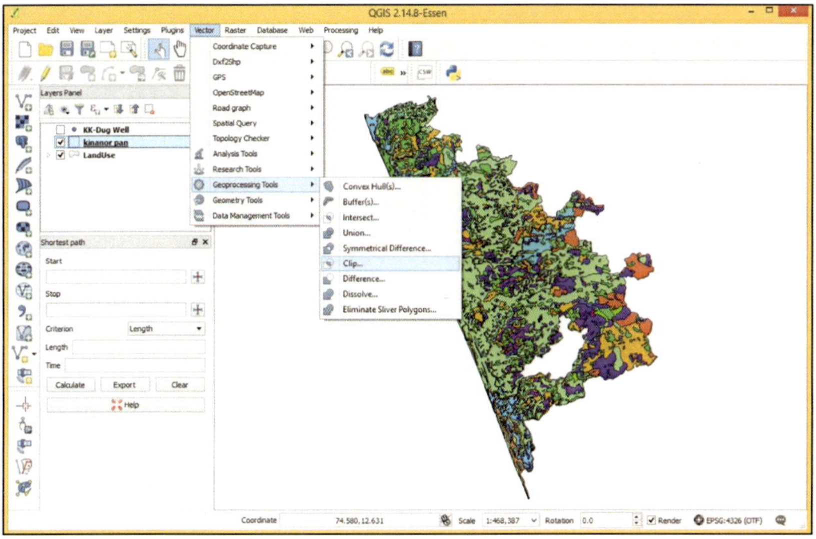

2. Select the appropriate Input vector layer and Clip layer

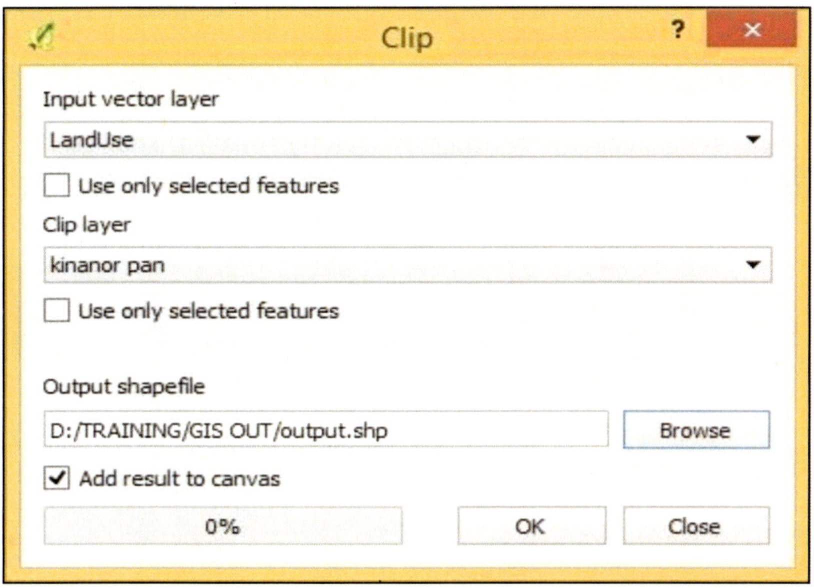

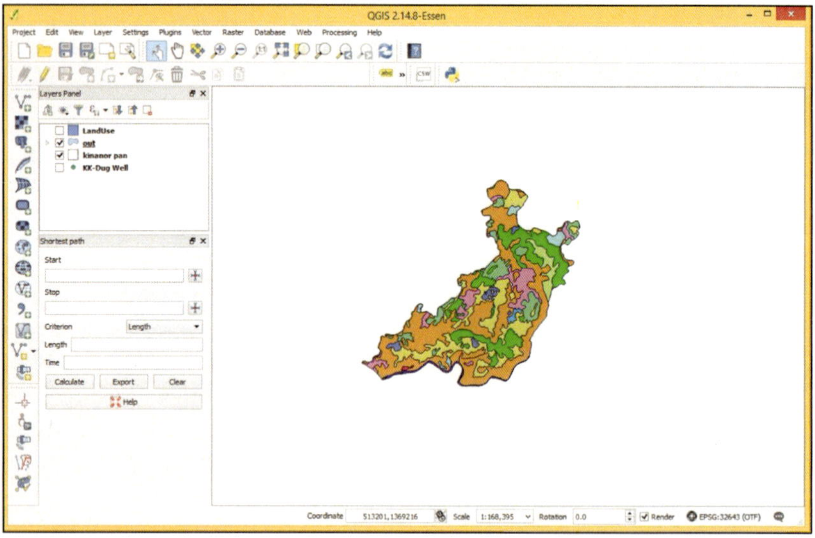

Interpolate

Theory

Interpolation is used when we collect data at a few locations and we have to mathematically predict the data for some unknown location/s. To interpolate a data, say rainfall data for a required area, the following steps can be used.

Procedure

1. With data ready, select Raster tab -Interpolation-Interpolation

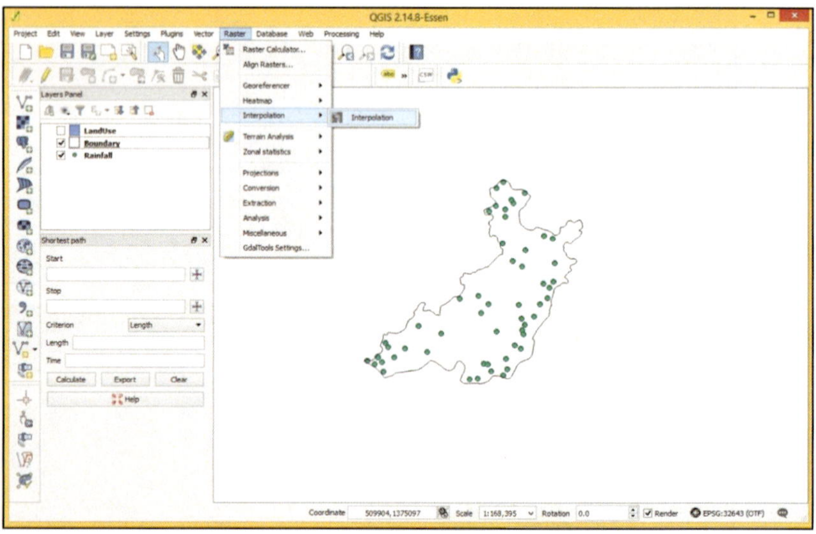

2. Select the Input layer and Interpolation attribute and click add tab.

   In the Output section, select the Interpolation method – "Triangular interpolation"(TIN) or "Inverse Distance Weighting" (IDW).

   Click on "Set to current extent" so that interpolation operates over entire boundary required.

   Save the Output file.

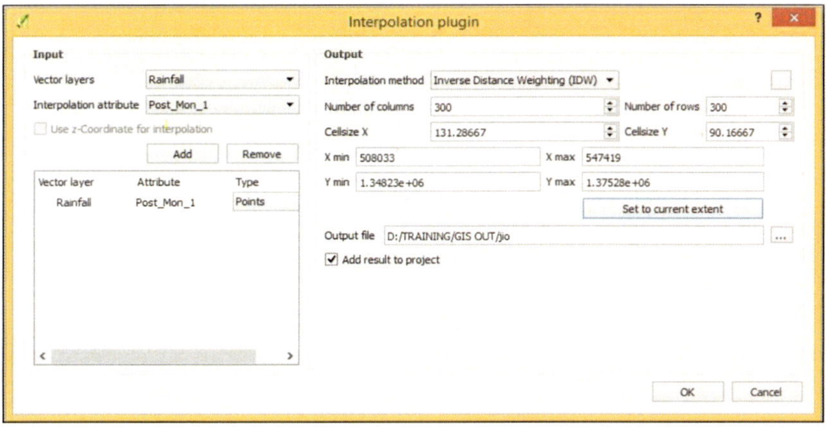

3. After interpolation, extract the boundary using Extraction Tool.

   Select Raster tab Extraction-Clipper

   Enter the Input file and save it to an output file.

   Select the Mask layer in Clipping mode.

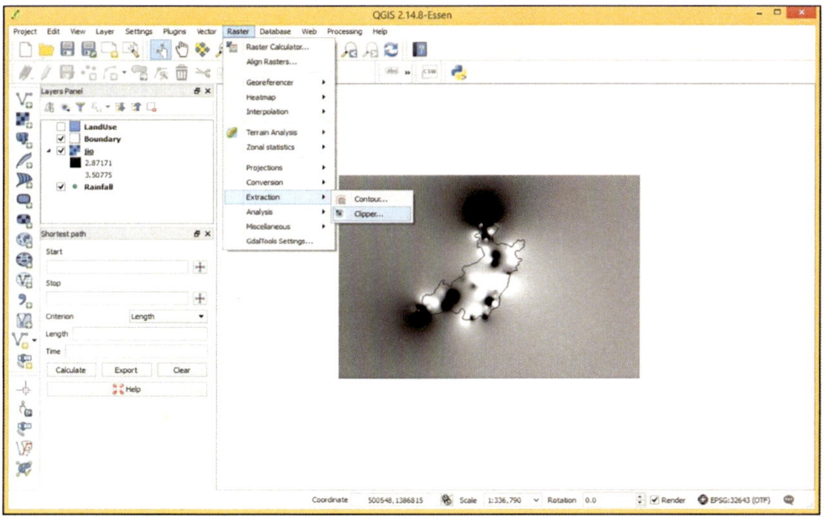

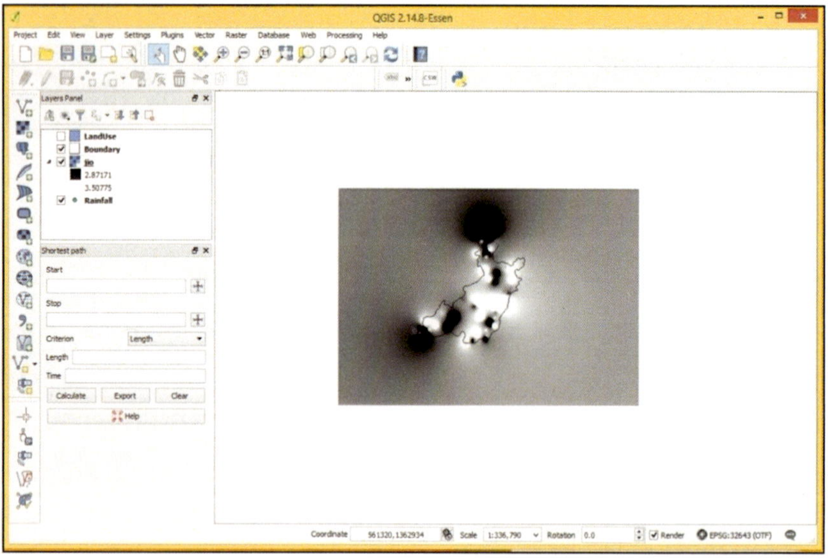

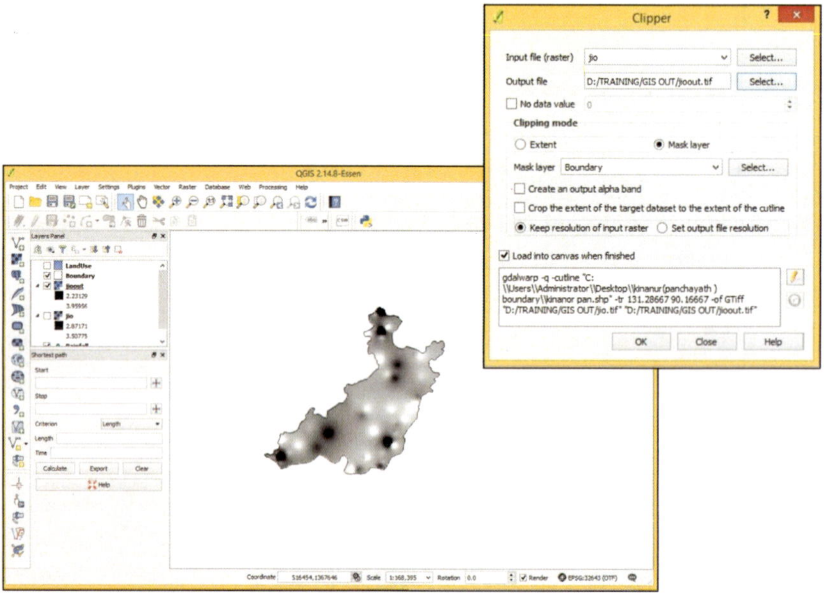

After Clipping, Layer properties like Color, Classification, etc. can be arranged.

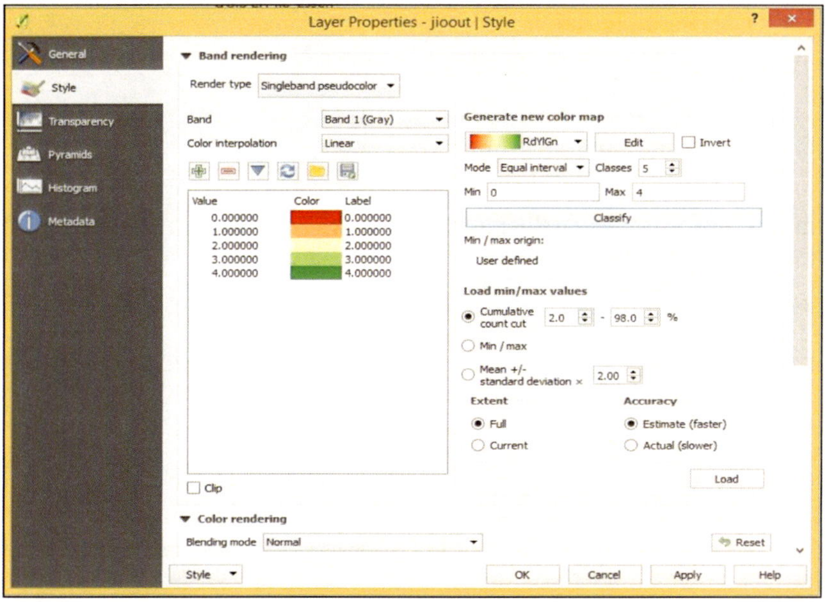
Layer Properties - jioout | Style
General
Style
Transparency
Pyramids
Histogram
Metadata
Band rendering
Render type Singleband pseudocolor
Band Band 1 (Gray)
Color interpolation Linear
Value Color Label
0.000000 0.000000
1.000000 1.000000
2.000000 2.000000
3.000000 3.000000
4.000000 4.000000
Clip
Generate new color map
RdYlGn Edit Invert
Mode Equal interval Classes 5
Min 0 Max 4
Classify
Min / max origin:
User defined
Load min/max values
Cumulative count cut 2.0 - 98.0 %
Min / max
Mean +/- standard deviation × 2.00
Extent
Full
Current
Accuracy
Estimate (faster)
Actual (slower)
Load
Color rendering
Blending mode Normal
Reset
Style
OK Cancel Apply Help

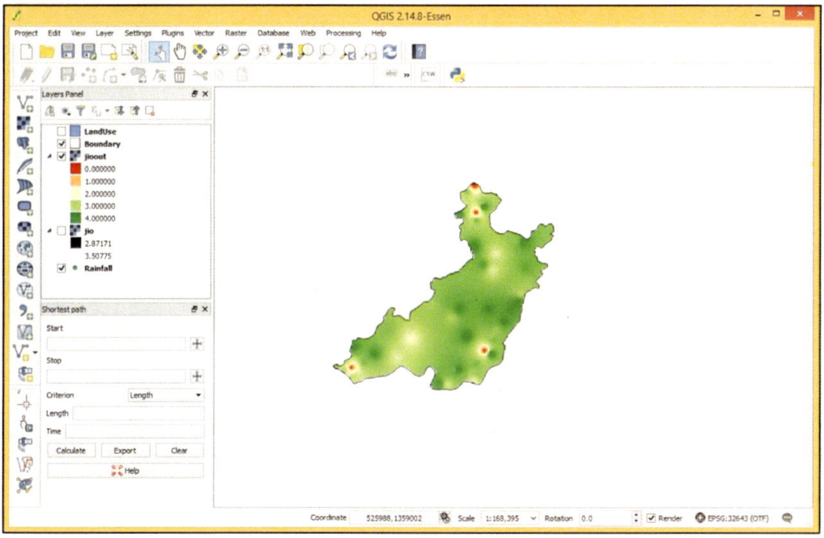
QGIS 2.14.8-Essen
Project Edit View Layer Settings Plugins Vector Raster Database Web Processing Help
Layers Panel
LandUse
Boundary
jioout
0.000000
1.000000
2.000000
3.000000
4.000000
jio
2.87171
3.50775
Rainfall
Shortest path
Start
Stop
Criterion Length
Length
Time
Calculate Export Clear
Help
Coordinate 525988,1359002
Scale 1:168,395
Rotation 0.0
Render
EPSG:32643 (OTF)

Value Tool

Theory

This tool is used to display the values in a table or plot from the visible raster layers at the current mouse position.

For example, to know the exact value of "depth to water table" or "rainfall in mm" of a particular location in an area (say, panchayath) during the pre-monsoon and post-monsoon period, proceed with the steps given below.

Procedure

1. The interpolated raster layer of pre-monsoon and post-monsoon is selected.

   If the "Value Tool" is not enabled, fetch it from the Plugins tab. On installing the "Value Tool," it will be displayed on the left side of the QGIS window.

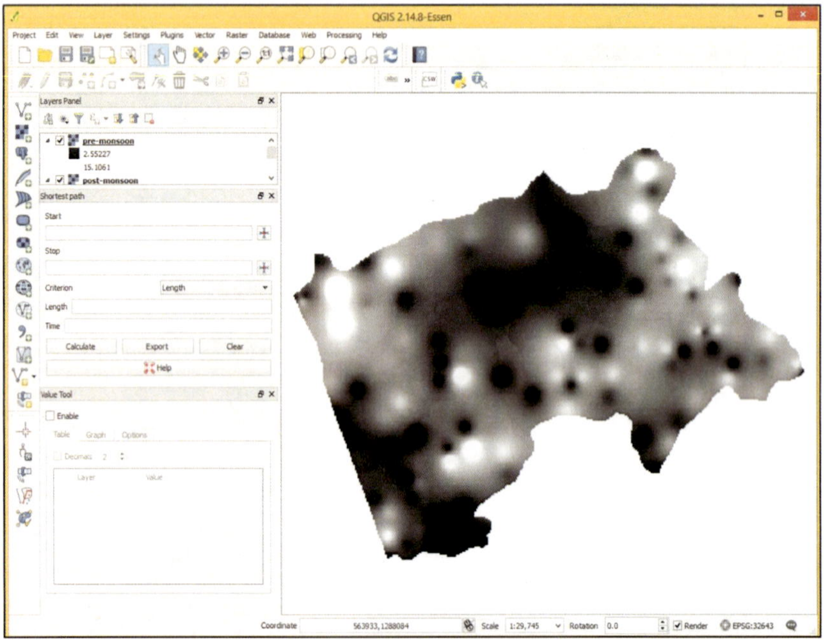

2. Enable the value tool box. Once the layer properties like Classification, color, etc. are arranged, just move the cursor to the required point in the area where the value of "depth to water table" or "rainfall" has to be known.

   The Value tool box will display the value of that particular point during the pre-monsoon and post-monsoon periods.

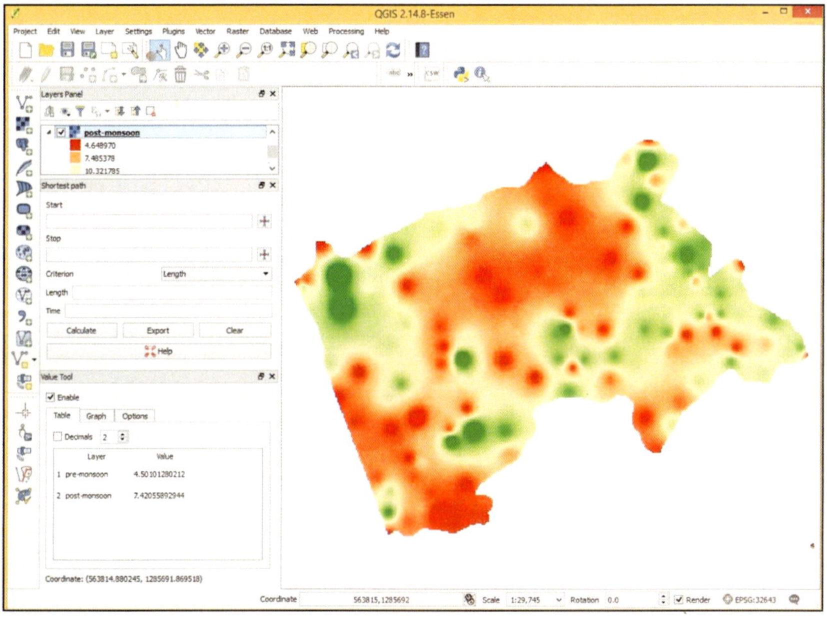

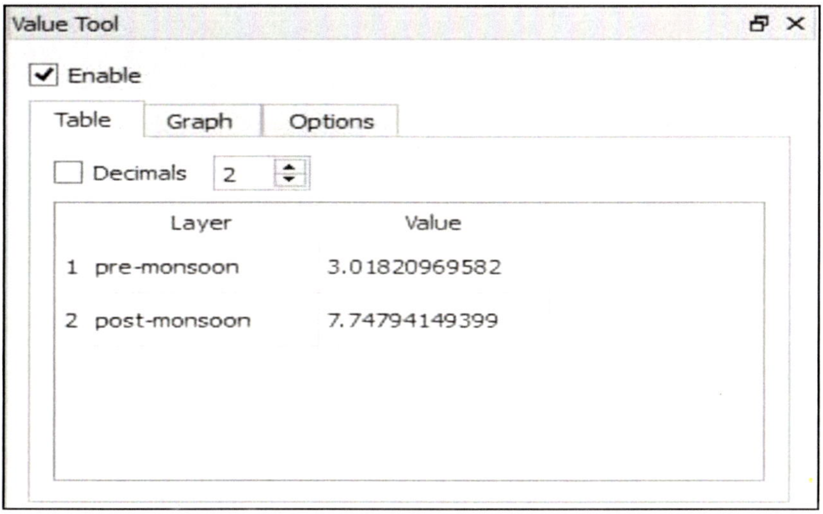

Split

Theory

This algorithm takes a vector layer and an attribute, and generates a set of vector layers in an output folder. Each of the layers created in that folder contains all features from the input layer with the same value for the specified attribute.

The number of files generated is equal to the number of different values found for the specified attribute.

Open a shape file (vector layer)

To split the shape file into different types mentioned in the attribute table, proceed with the following steps.

Procedure

1. From the Vector tab, select Data Management Tools and from there, select Split vector layer.

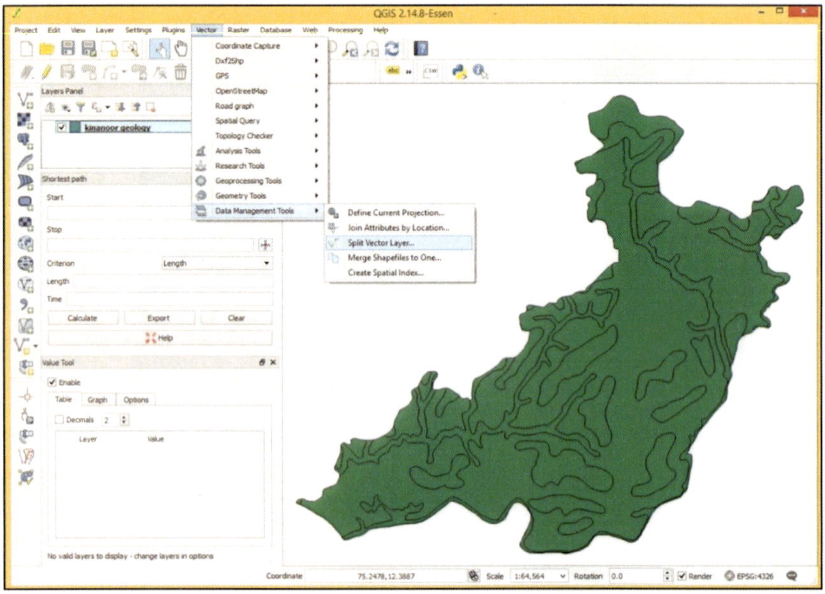

2. Select the Unique ID field and save the Output Shape file in a folder

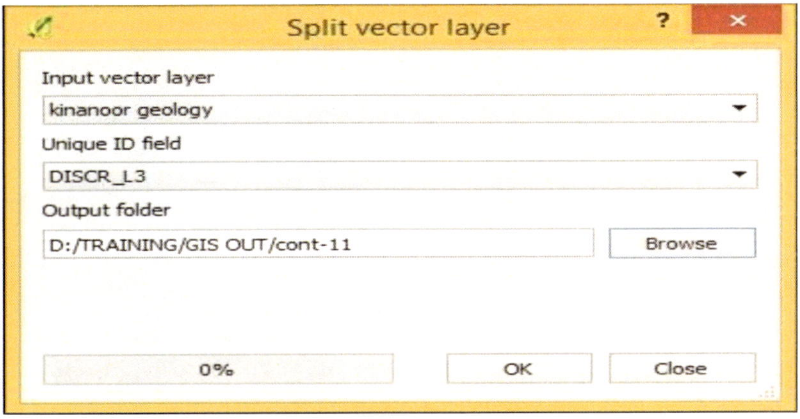

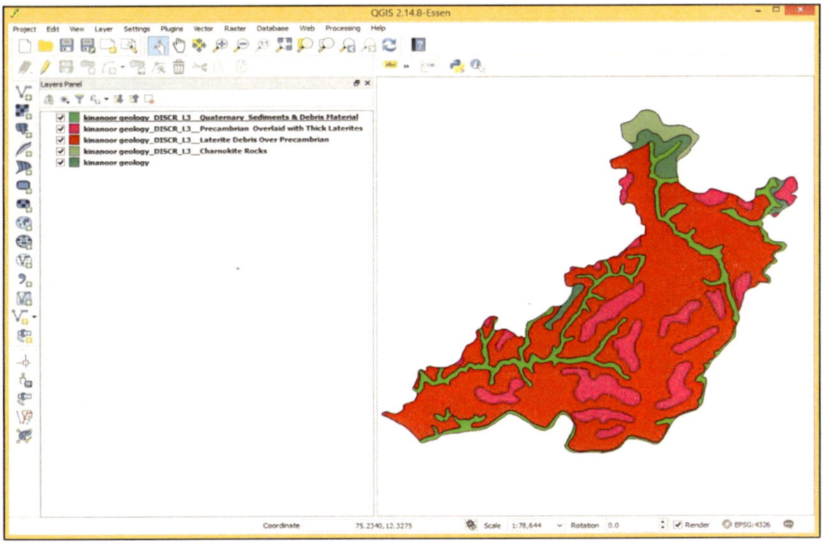

3. Likewise, “Merge Shapefile to One” can also be done

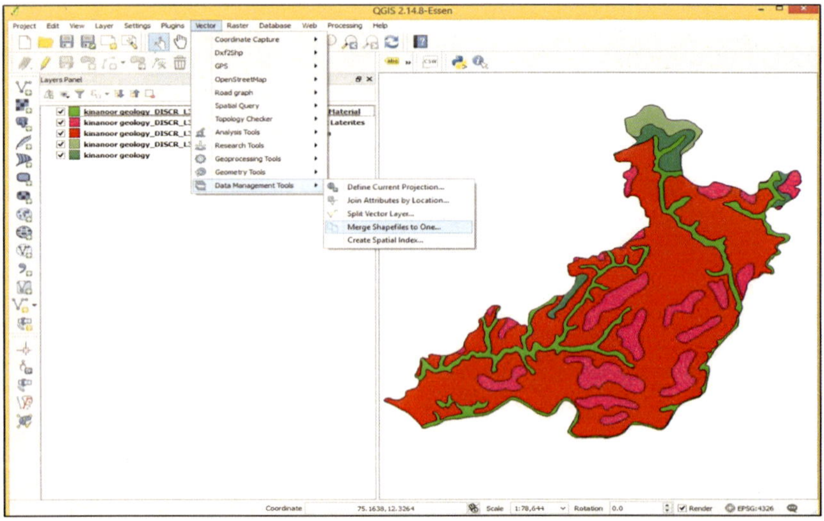

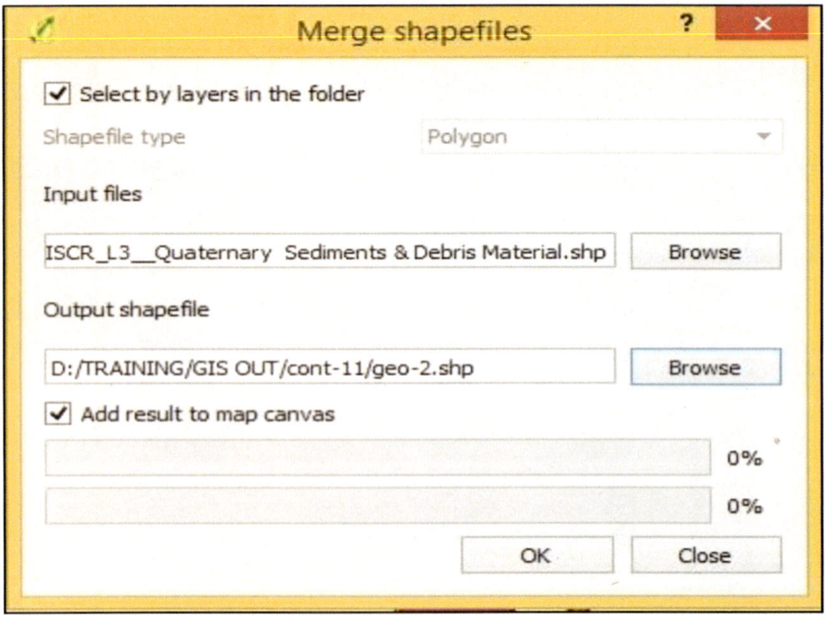

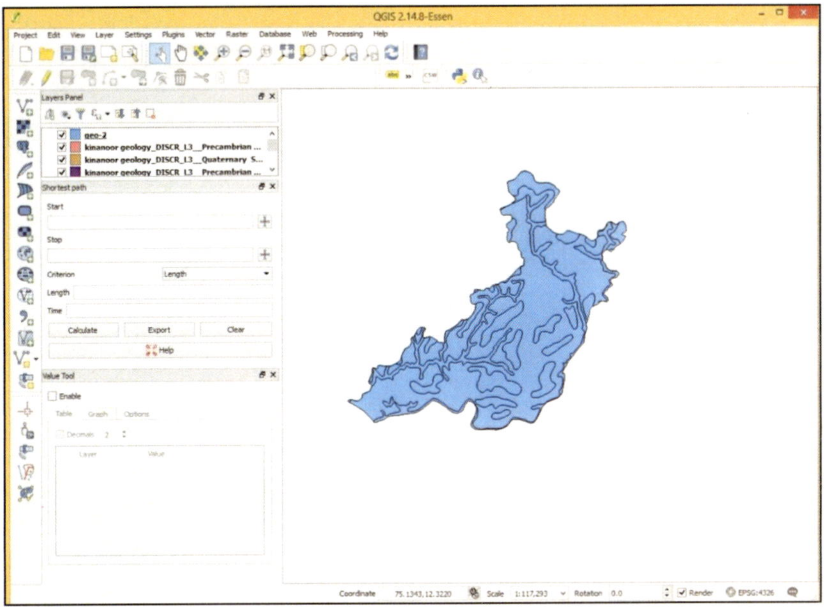

Spatial Join

Theory

Spatial Join is a classic GIS problem—transferring attributes from one layer to another based on their spatial relationship. In QGIS, this functionality is available through the "Join Attributes by Location tool."

Procedure

1. Go to Layer tab → Add Vector Layers

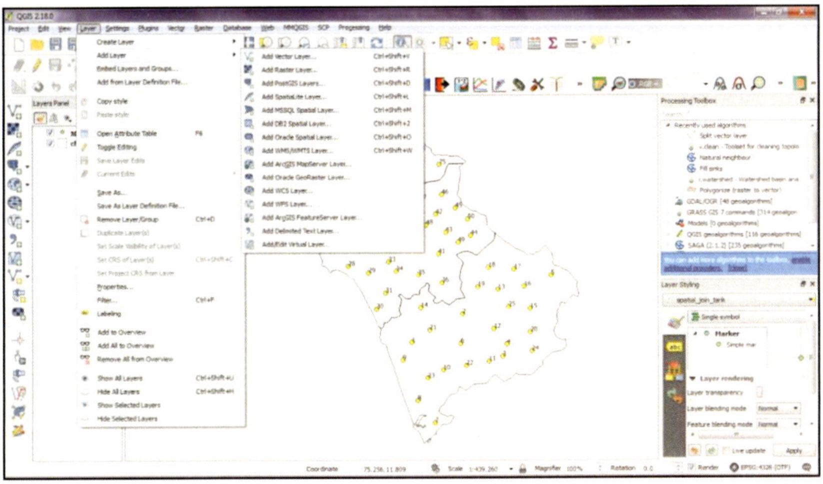

2. Examine the attribute available for each feature, right click on Target layer select "Open Attribute Table"

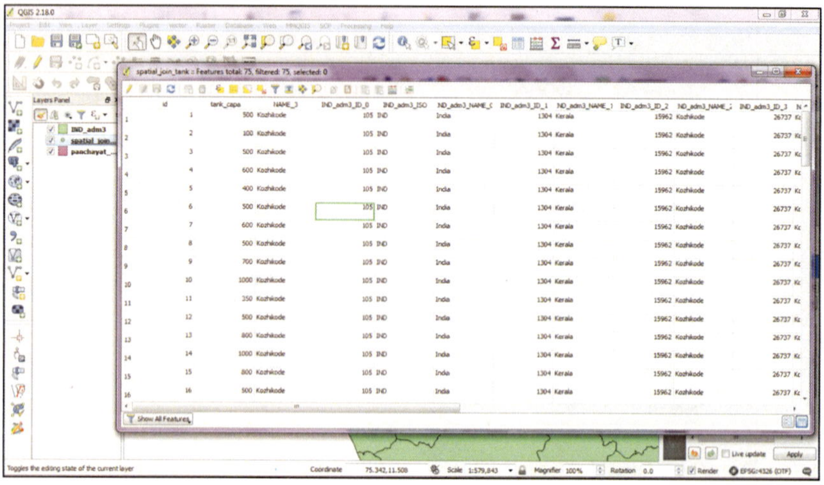

3. Go to Vector tab → Data Management Tools → Join Attribute by Location and fill the settings → Run

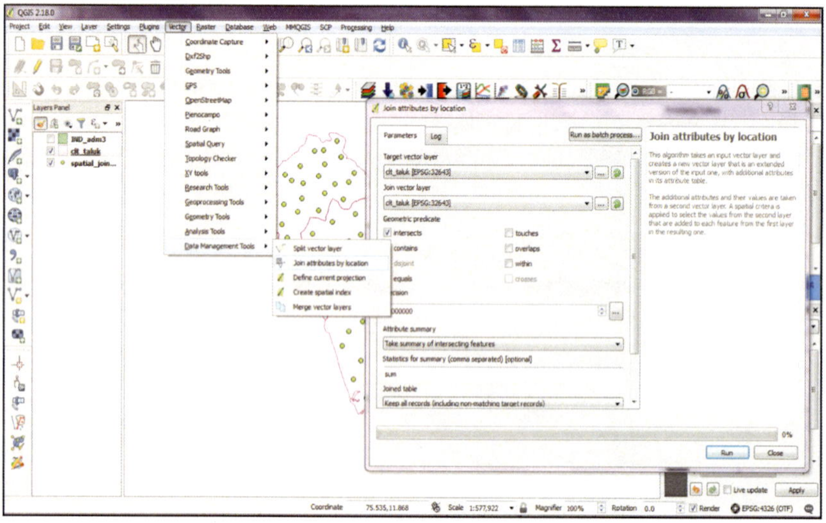

4. After getting output (Joined layer) → click "Start Editing mode " on the attribute tableand then use "Delete field " option to delete unwanted columns.

   Then Spatial join of attributes is finished.

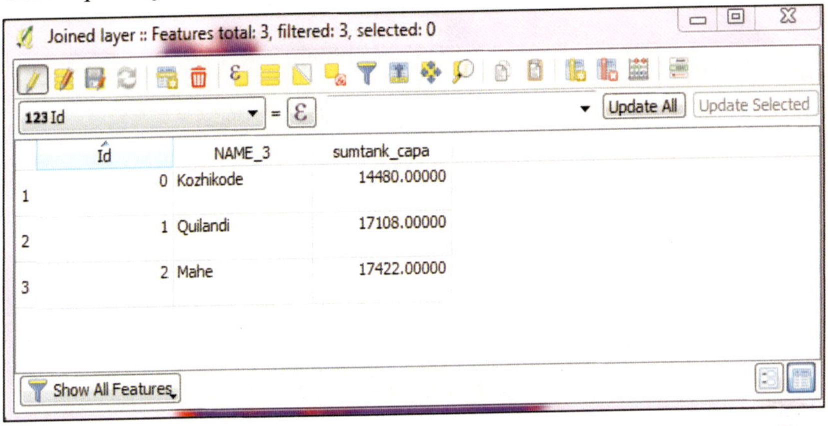

Generation of Slope

Theory

Slope is the measure of steepness or the degree of inclination of a feature relative to the horizontal plane. It is crucial to do a study on the slope of the area of interest (aoi) to reduce construction costs, minimize risks from natural hazards such as flooding and landslides, and minimize the impacts of proposed development on natural resources such as soils, vegetation, water systems, etc. In this chapter, we will be learning how to generate a slope map from a digital elevation model (DEM).

Procedure

1. Go to *Layer* → *Add layer* → *Add raster layer*. To clip the required study area, add the shapefile of the study area and go to *Raster* → *Extraction* → *Clip raster by mask layer*.

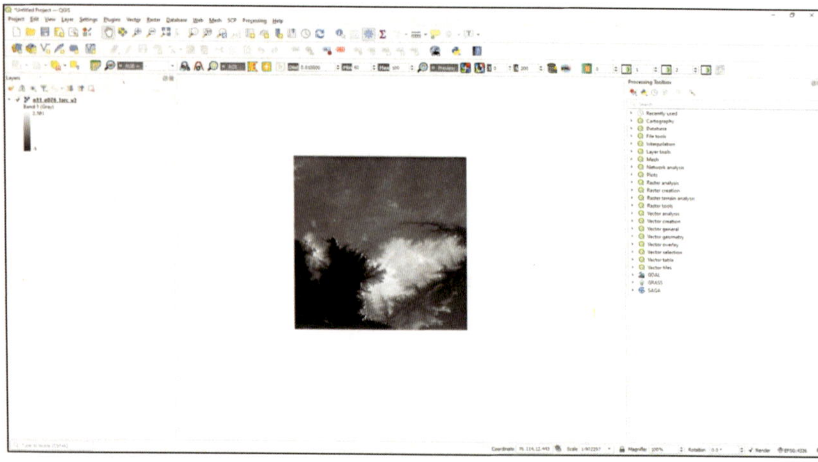

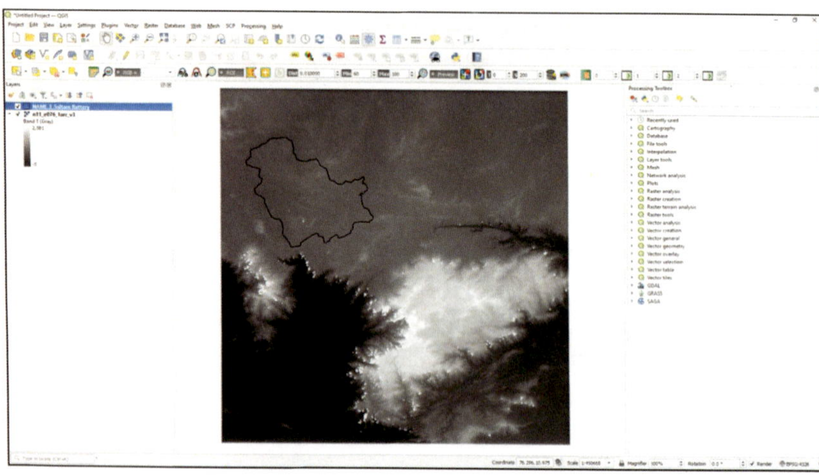

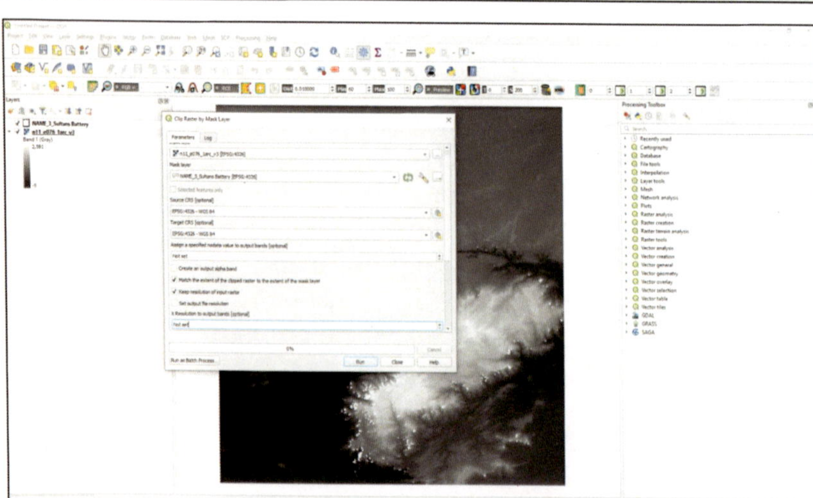

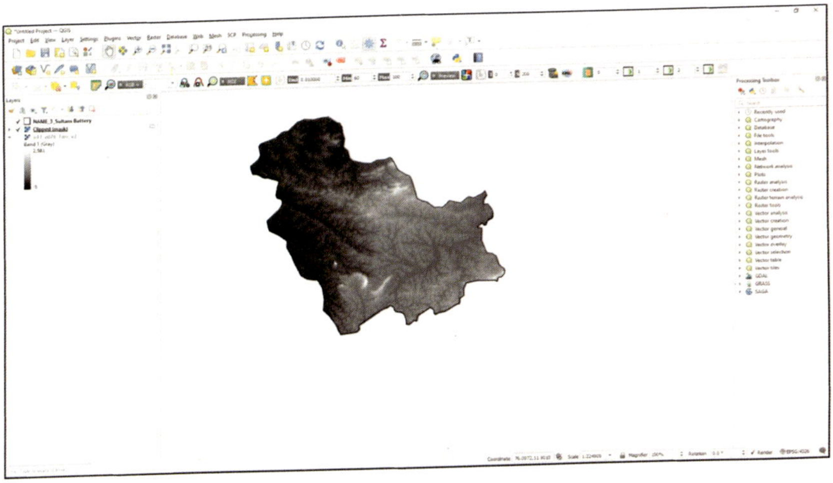

2. Go to *Raster* → *Projections* → *Warp (Reproject)*

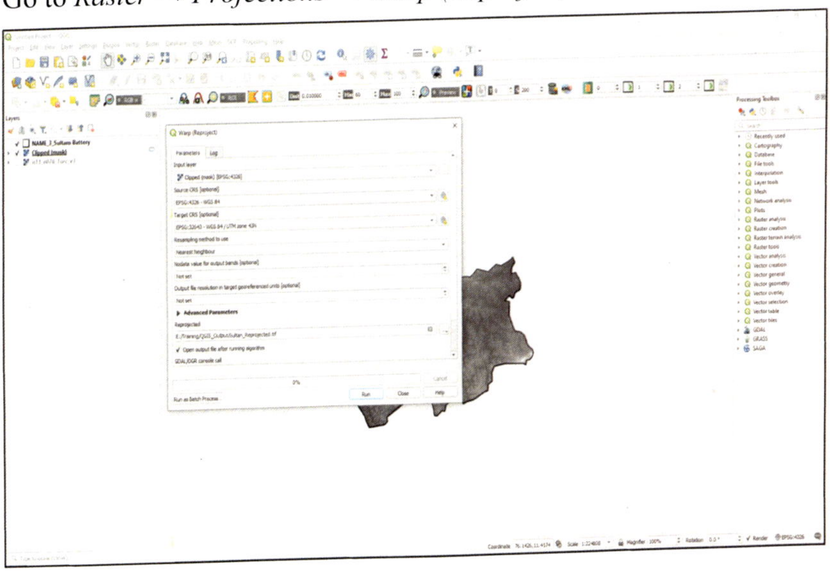

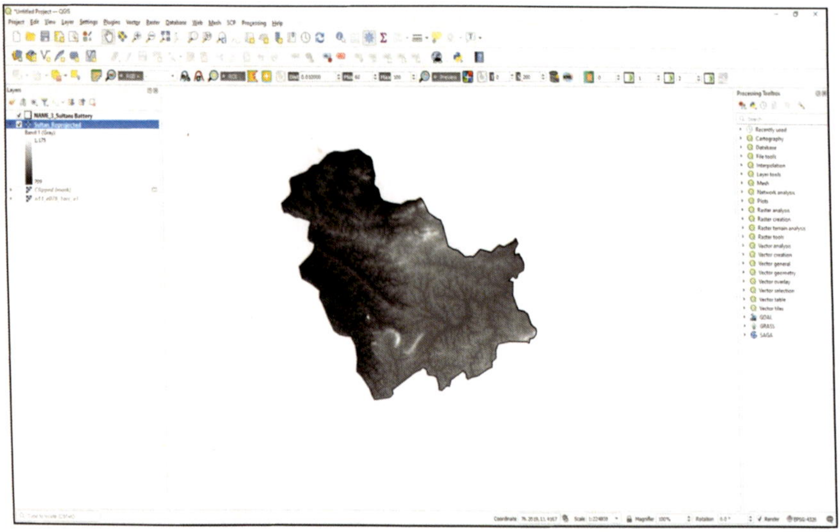

3. Go to *Raster* → *Analysis* → *Slope*. Provide the input layer and click *Run*.

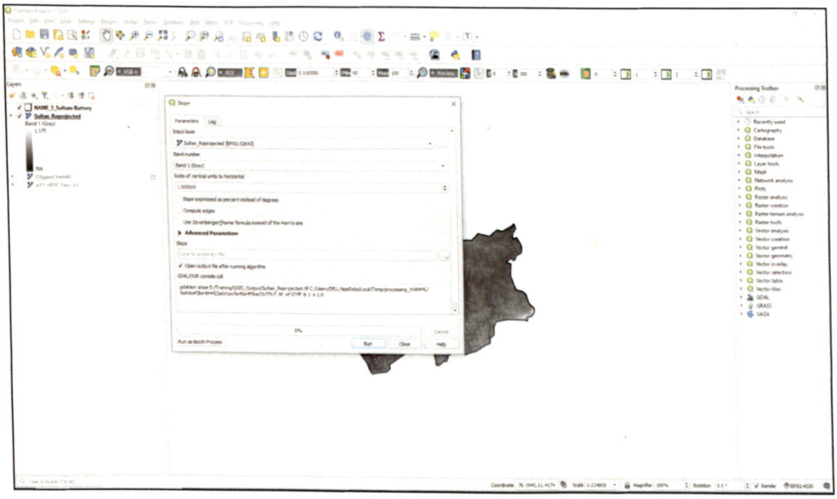

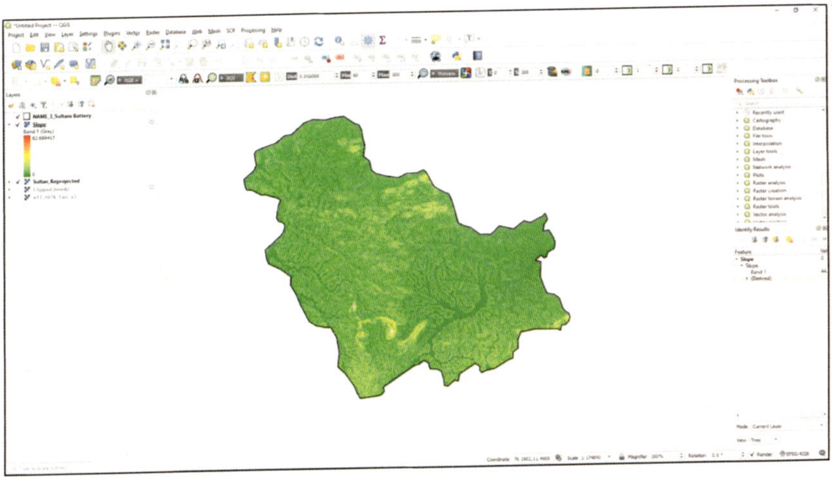

Generation of Contours

Theory

Contours are imaginary lines connecting points having the same elevation on the surface of the land above or below mean sea level. The contour map provides important information regarding the nature of the terrain. Contour maps can be useful for site selection, determining the catchment area of a drainage basin, determining station inter visibility, etc. With the help of contours, it is possible to depict the height and shape of mountains, the depths of the ocean bottom, and the steepness of the slopes. In this chapter, we will learn how to extract contours from a digital elevation model (DEM).

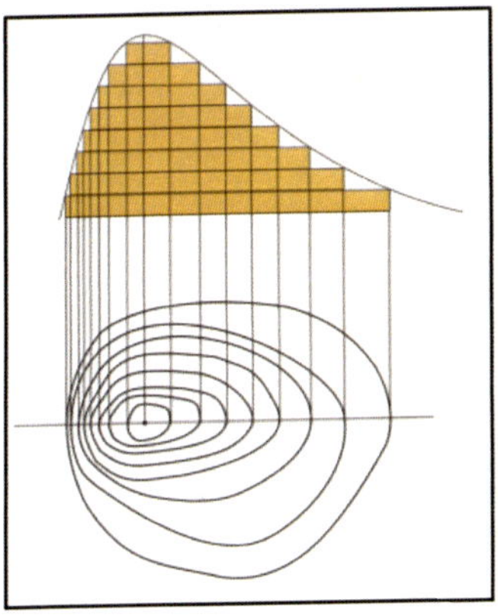

(Source: en.wikipedia.org)

Procedure

1. Go to *Layer* → *Add layer* → *Add raster layer*. To clip the required study area, add the shapefile of the study area and go to *Raster* → *Extraction* → *Clip raster by mask layer*.

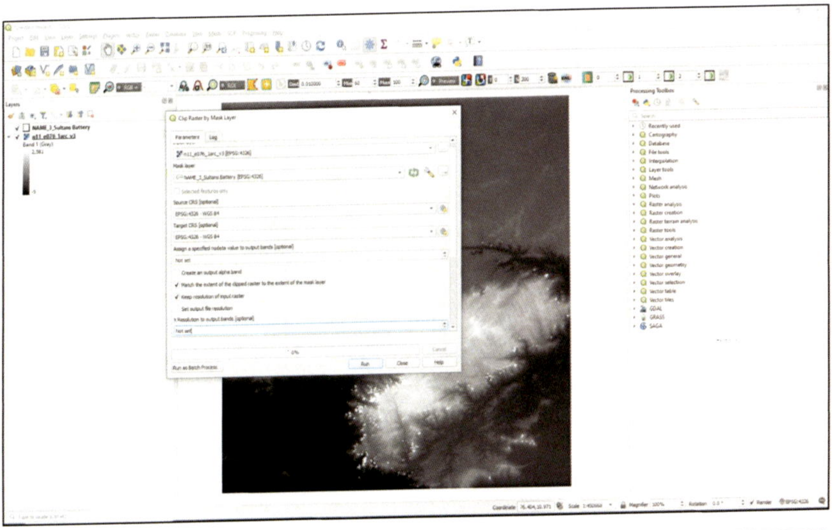

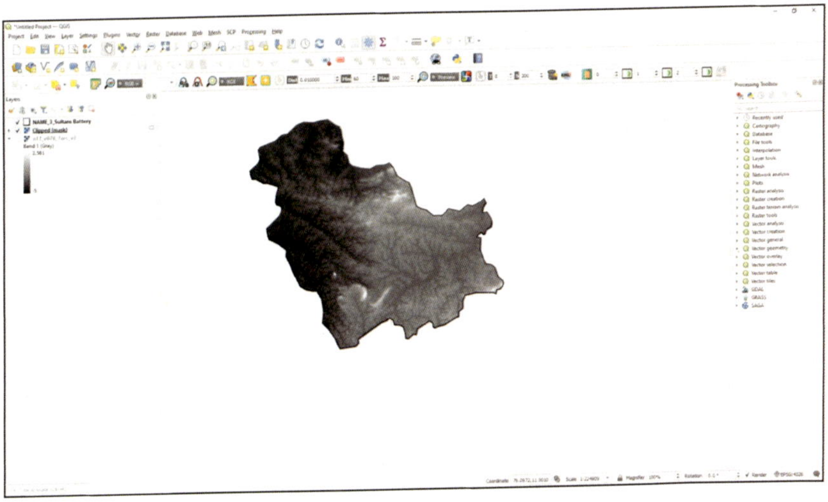

2. Go to *Raster → Projections → Warp (Reproject).*

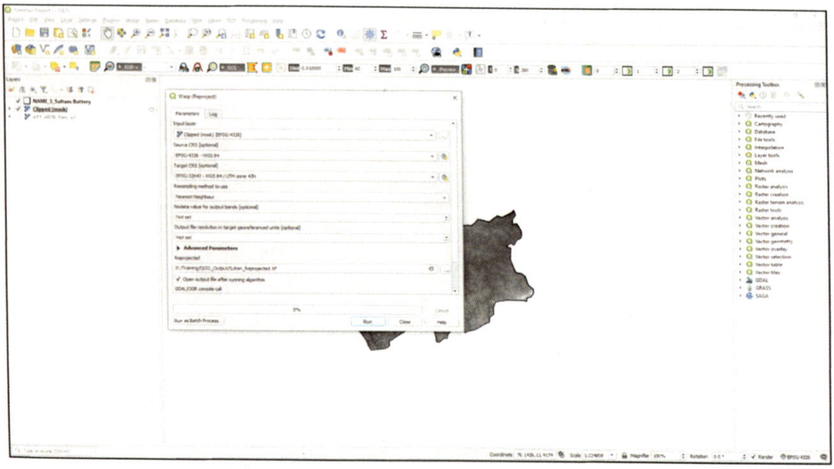

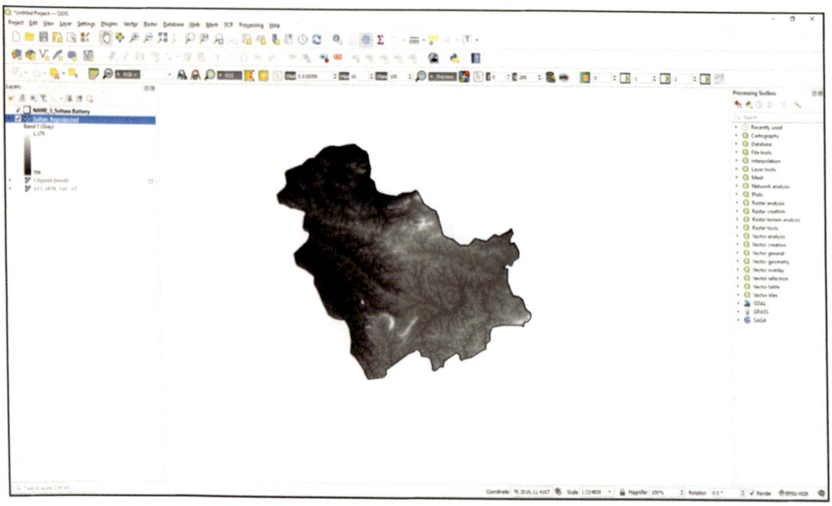

3. Go to *Raster* → *Extraction* → *Contour*. Provide the input layer and the interval between the contour lines and click *Run*.

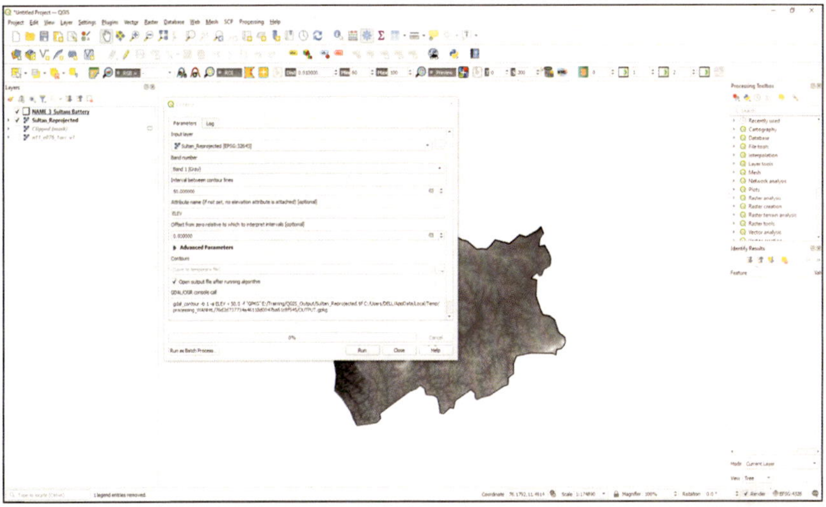

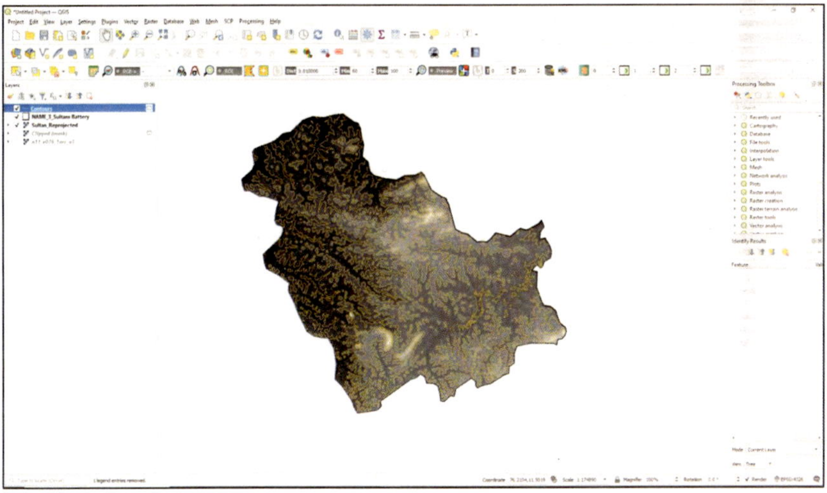

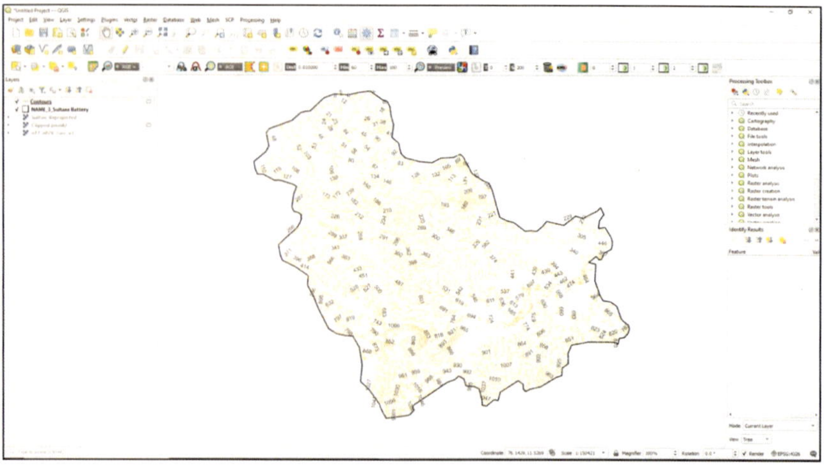

Map Composition

Theory

Map composition helps in bringing together various symbols on the map so that the map can communicate spatial, topographic, and geographic information effectively. The user can adjust the size, position, and properties of each symbol to create the layout. The result can be exported as an image, PDF, or SVG.

Procedure

1. Start a new project and go to *Project → New print layout*. Provide a unique title to the layout and click *OK*. A new dialogue box pops up.

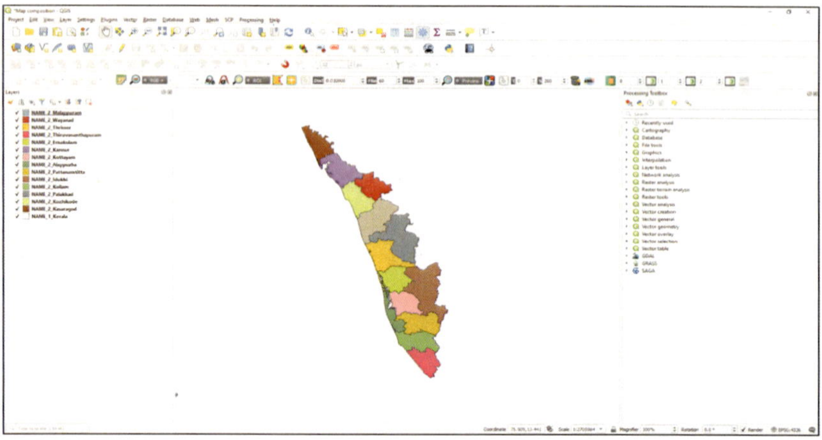

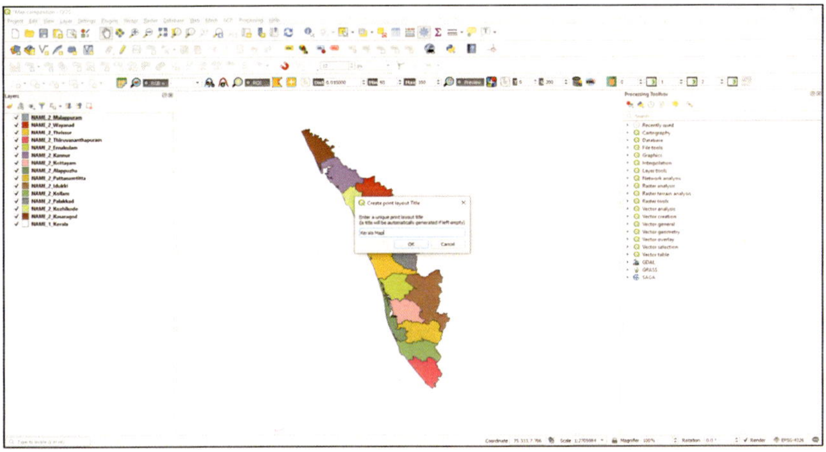

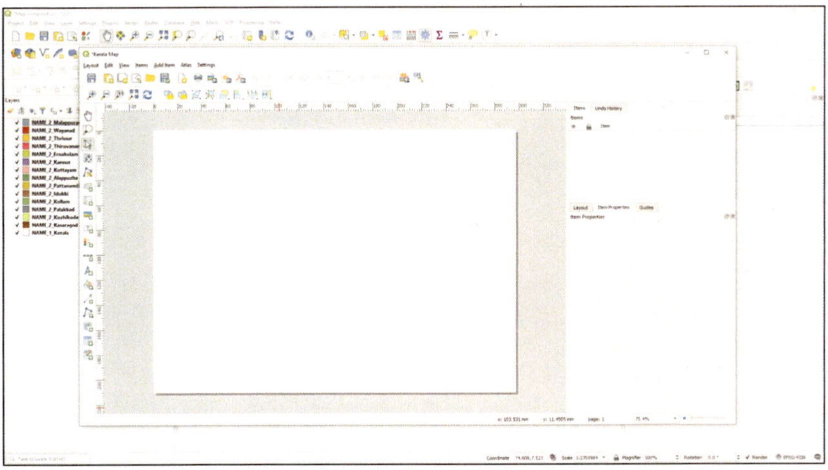

2. Click on *Add Map icon* [ ] in the Toolbox and with the help of the cursor, draw a box. The layers that were already present on the project will appear in this layout.

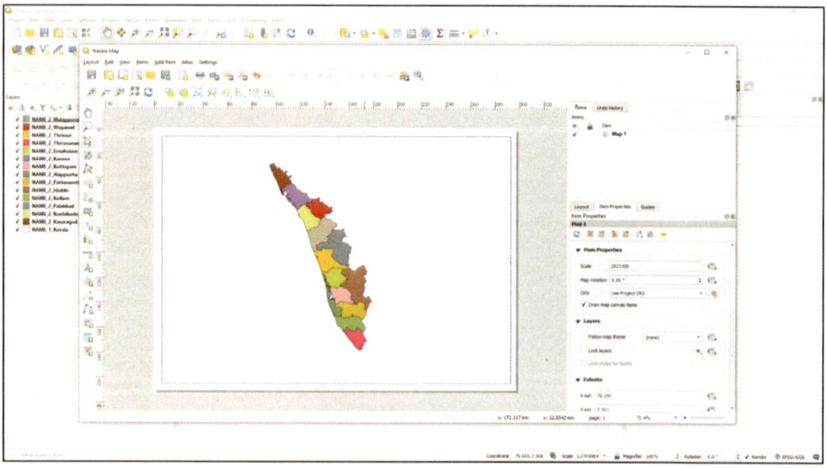

3. To add coordinates to the map, go to *Grids* in *Item Properties*. Click on *Add a new grid* [] and click on *Modify Grid*. Select the map grid properties according to your requirements.

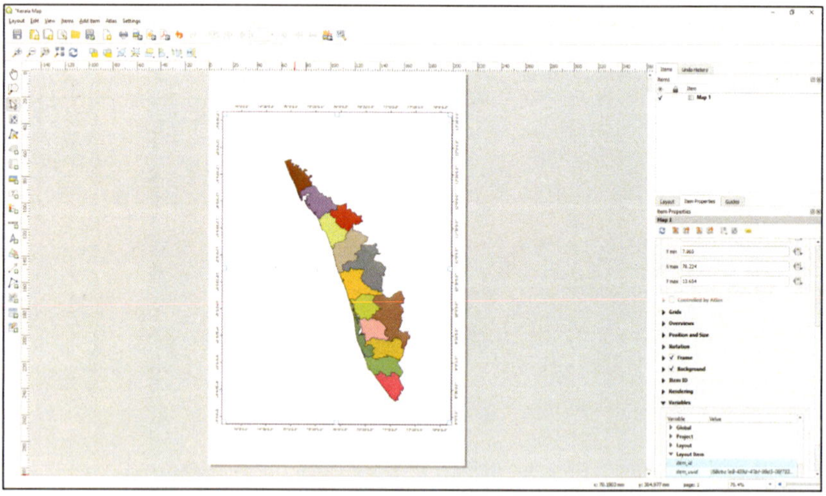

4. From the Toolbox, select *Add North Arrow icon* []. Drag a box and add North arrow to the map.

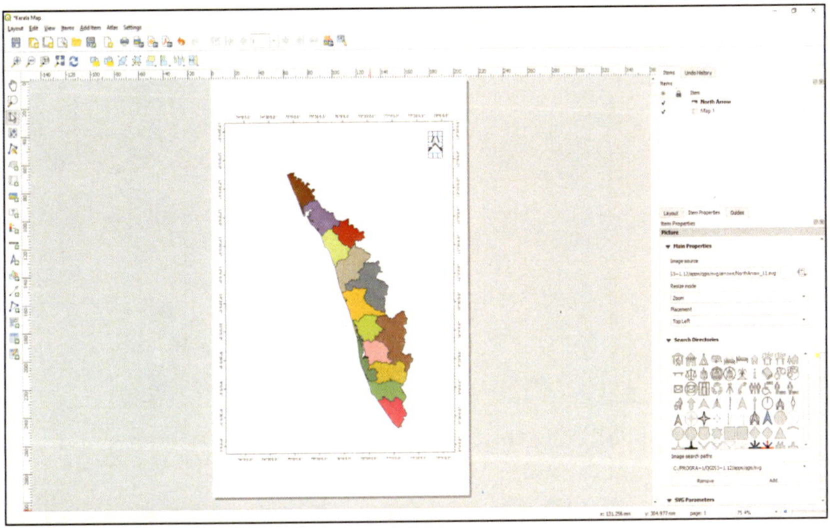

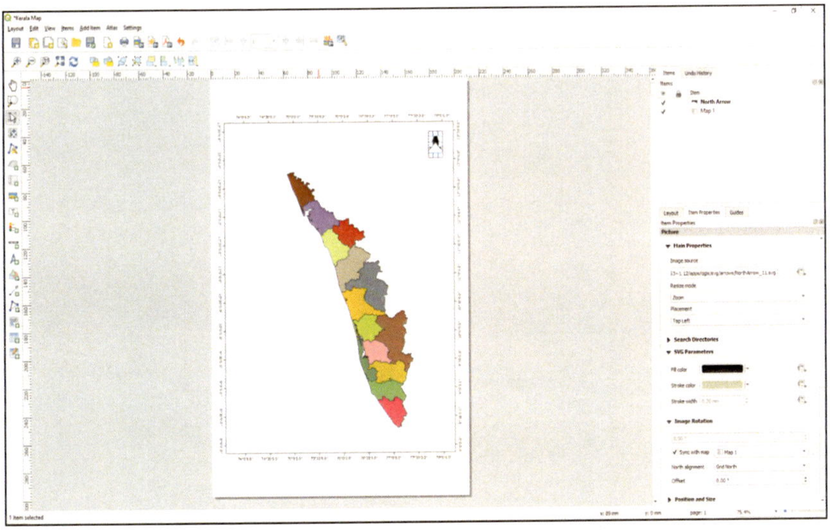

5. From the Toolbox, select *Add Scale Bar* [ ]. Drag and draw a box to add a scale bar to the map.

6. From the toolbox, select *Add Label icon* [ ]. Drag and draw to add a title bar and provide a name to the map.

7. From the Toolbox, select *Add Legend icon* []. Drag and draw a box and the layers that were already present in the project will appear in the legend box. Edit and keep only the required names in the legend box.

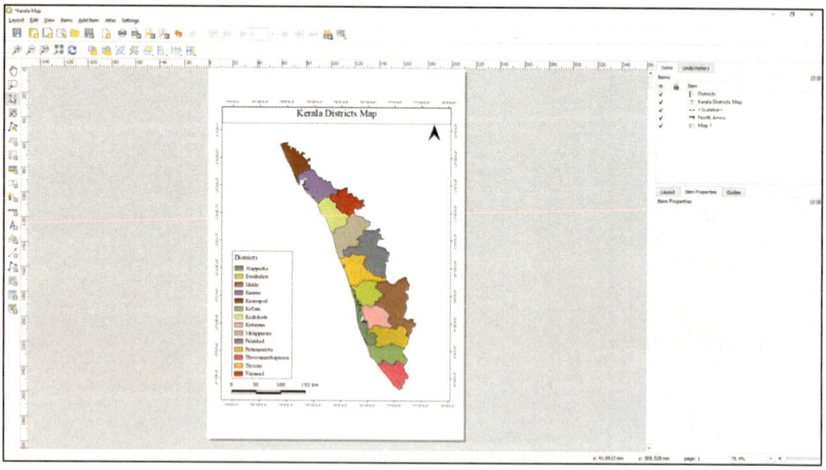

8. To save this map, go to *Layout* → *Export as Image*. Provide a name and click *Save*.

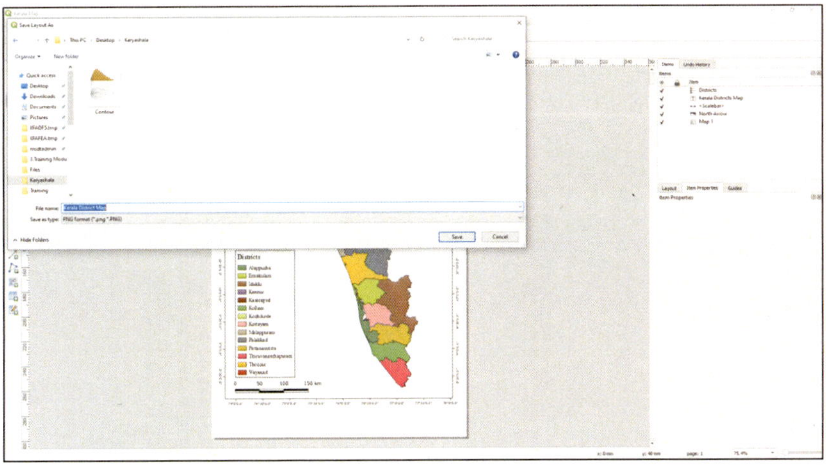

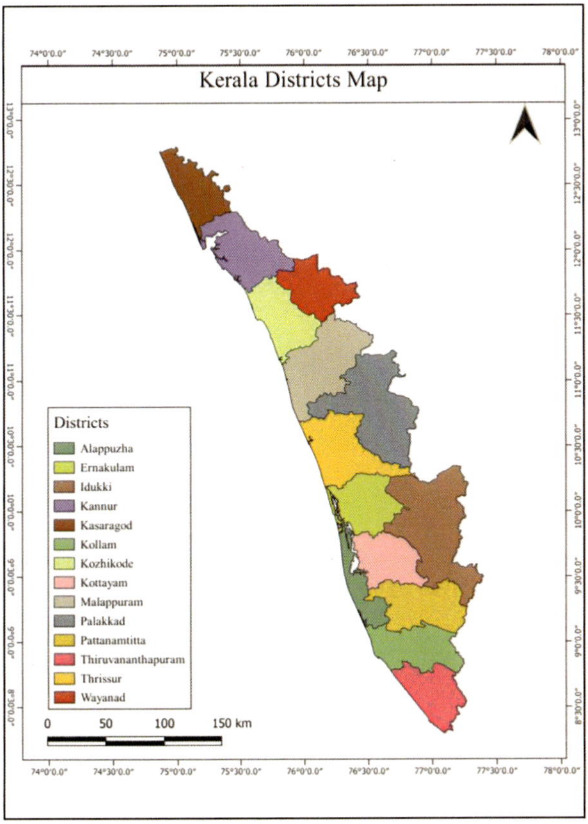

Import GPX File from GPS to QGIS

Procedure

1. The data collected using GPS (Waypoints, Tracks, etc.) will be in.gpx format which is a vector file. In this section we will learn how to access these points in GIS system and visualize, map, etc.

   Go to "Vector tab" → select GPS → GPS Tools

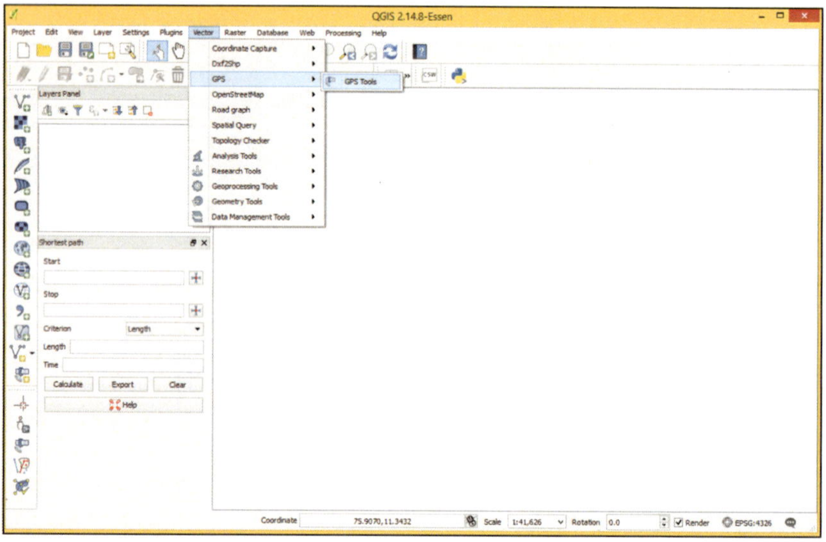

3. Select the required GPS file from Load GPX file

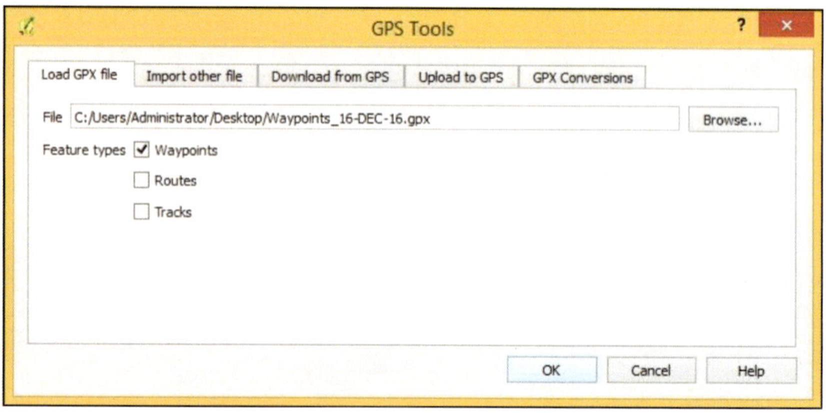

Linking Google Earth in QGIS

Procedure

1. Switch on the internet connection in the desktop/laptop
2. In Plugins tab, select Manage and Install Plugins

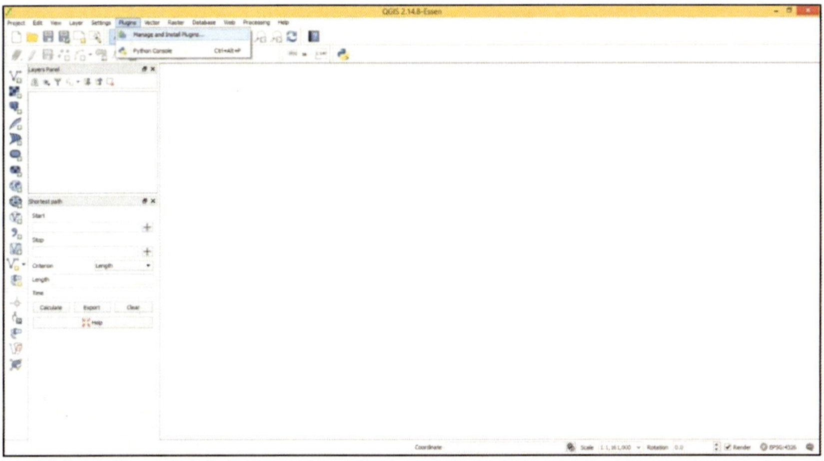

3. On the left, you see the list of all plugins available for your QGIS, both installed and available for download. Some plugins come with your QGIS installation while most of them are made available via the plugin repositories.

   You can temporarily enable or disable a plugin. To *enable* or *disable* a plugin, click its checkbox or double-click its name.

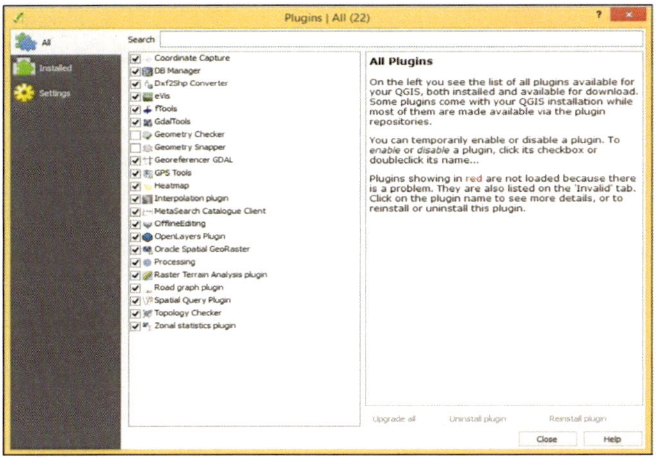

Enable "OpenLayers Plugin"(version 1.4.1)

Using this Plugin, one can open Google Maps, Bing Maps, OpenStreetMap layers, and more in QGIS.

4. To open Google Map, in Web tab, select OpenLayers plugin and from there, select the required Google Maps

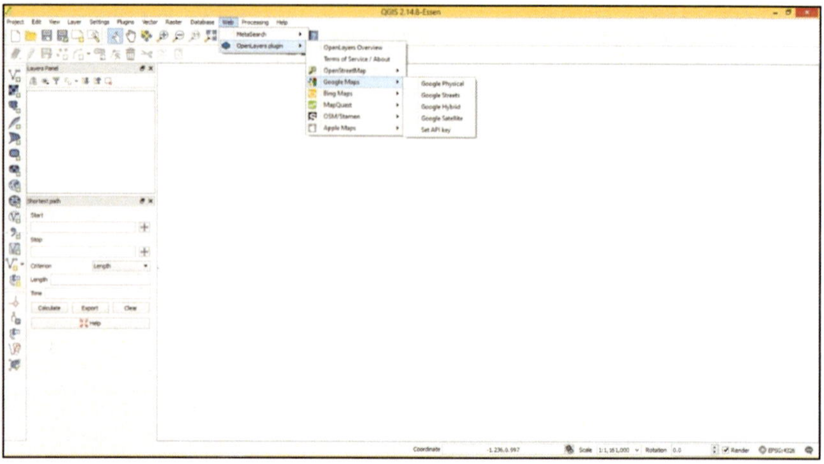

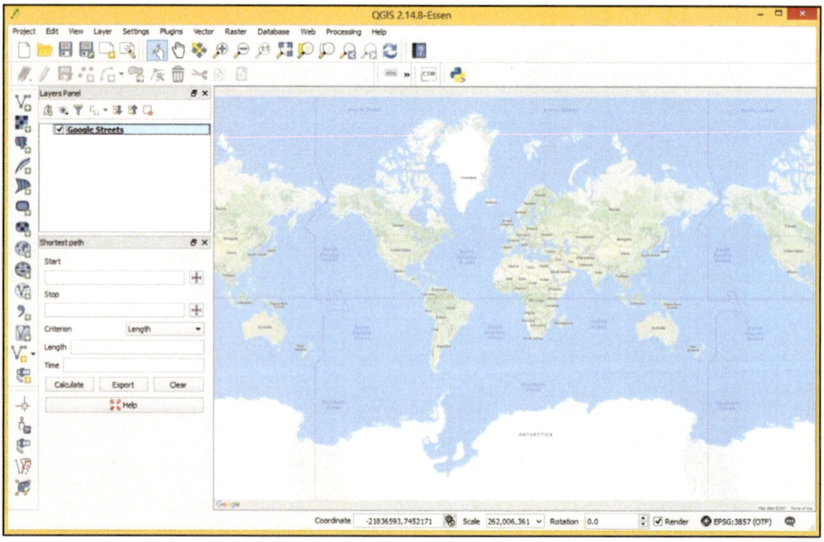

Web Map Services (WMS) in QGIS

Theory

Web Map Service Interface Standard (WMS) provides a simple HTTP interface for requesting geo-registered map images from one or more distributed geospatial databases. A WMS request defines the geographic layer(s) and area of interest to be processed. The response to the request is one or more geo-registered map images (returned as JPEG, PNG, etc.) that can be displayed in a browser application. The interface also supports the ability to specify whether the returned images should be transparent so that layers from multiple servers can be combined or not.

Procedure

1. Open QGIS and go to Layer tab → Add Layer → Select "Add WMS/ WMTS Layer... 

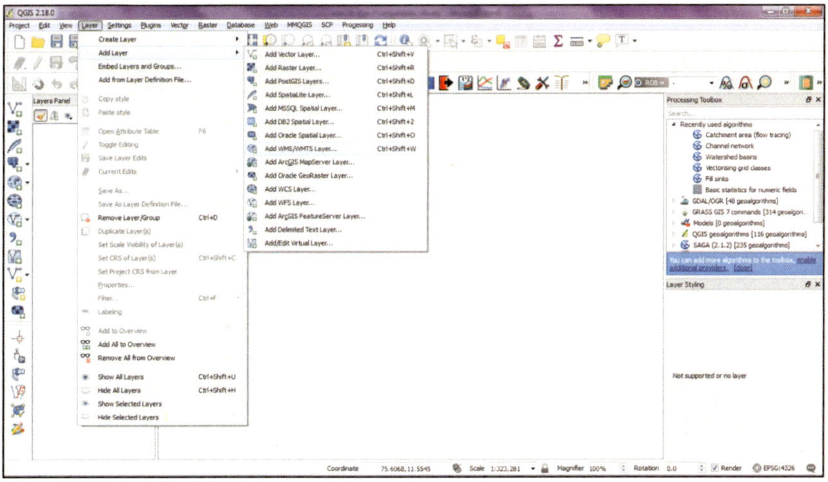

2. Click on "New" and then add Name & URL → OK

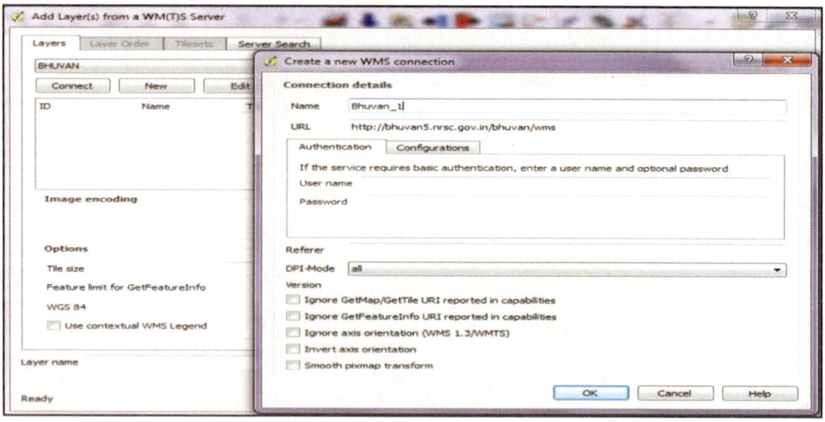

3. Click "Connect" and then Select the Layer → PNG → Add

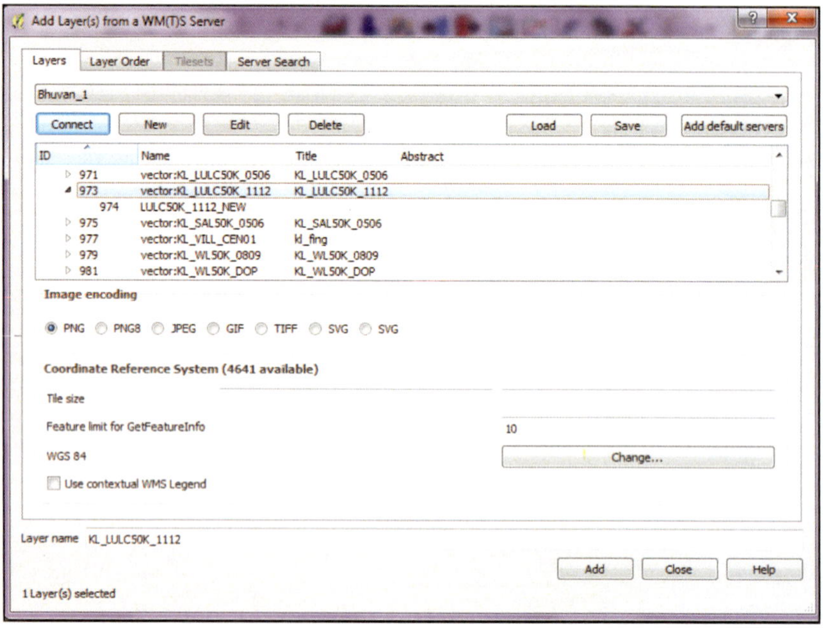

4. Final output

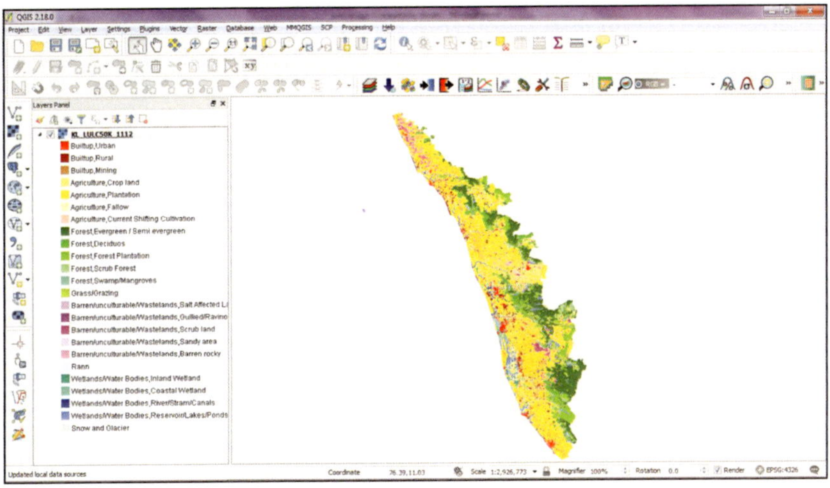

NDVI Generation

Theory

Vegetation indices combine surface reflectance at two or more wavelengths to emphasize a specific vegetation characteristic. The reflectance properties of vegetation are used to derive the vegetation index. The Normalized Difference Vegetation Index (NDVI) is a numerical indicator that utilizes the visible and near-infrared (NIR) regions of the electromagnetic spectrum. It is used to examine remote sensing measurements and determine whether or not the subject under observation has live green vegetation. NDVI has widespread usage in vegetative studies as it can be used to calculate crop yields, pasture performance, etc. The NDVI index is often related to parameters such as the percentage of ground cover, photosynthetic activity of the plant, leaf area index, and amount of biomass. NDVI is estimated using the formula:

$$NDVI = \frac{(NIR - Red)}{(NIR + Red)}$$

Procedure

1. Add red and NIR bands of the satellite image to the QGIS Project. For Landsat 8, band 4 is red and band 5 is NIR.

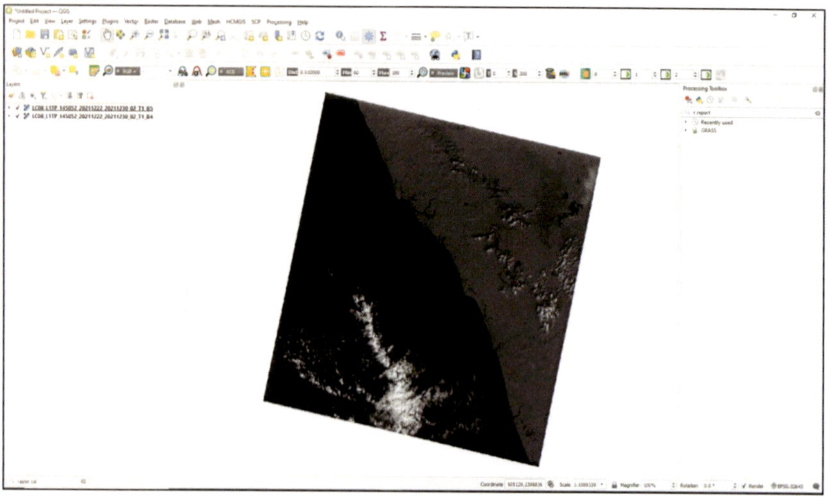

2. In the search option, search for Raster Calculator and select QGIS raster calculator.

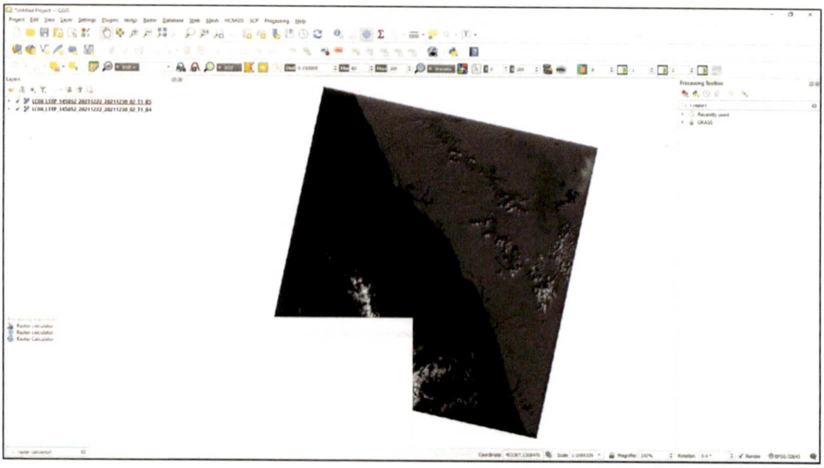

3. NDVI formula is to be written on the white box titled "Expression". After writing the expression, select an output option and then click *Run*.

$$NDVI = \frac{NIR - Red}{NIR + Red}$$

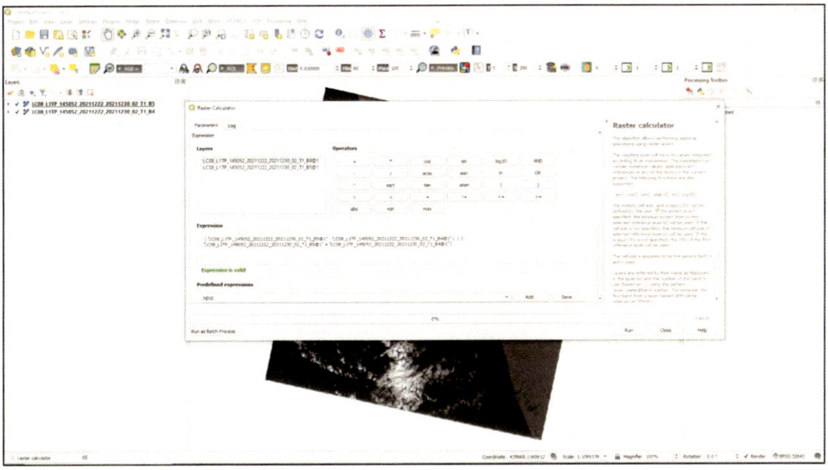

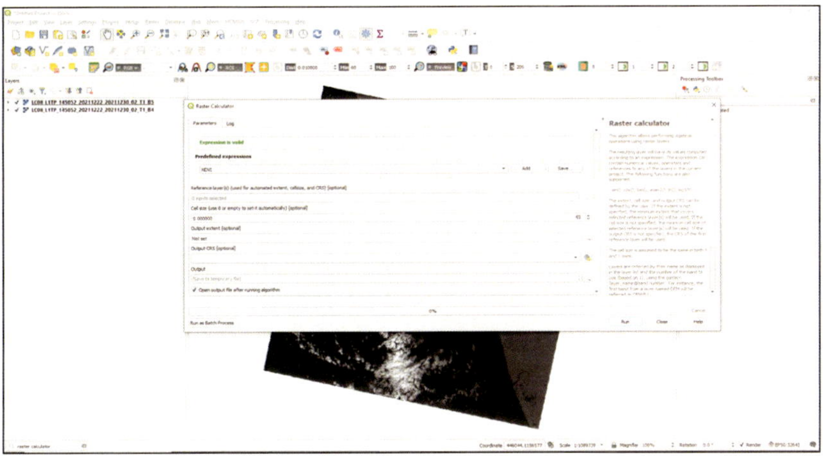

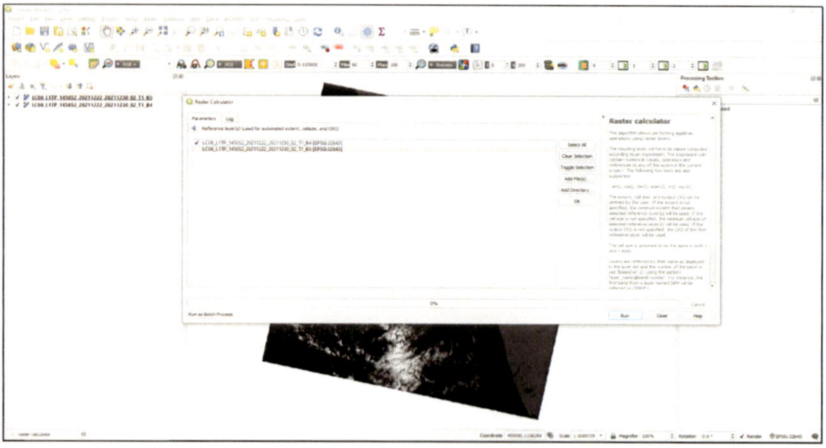
Raster calculator

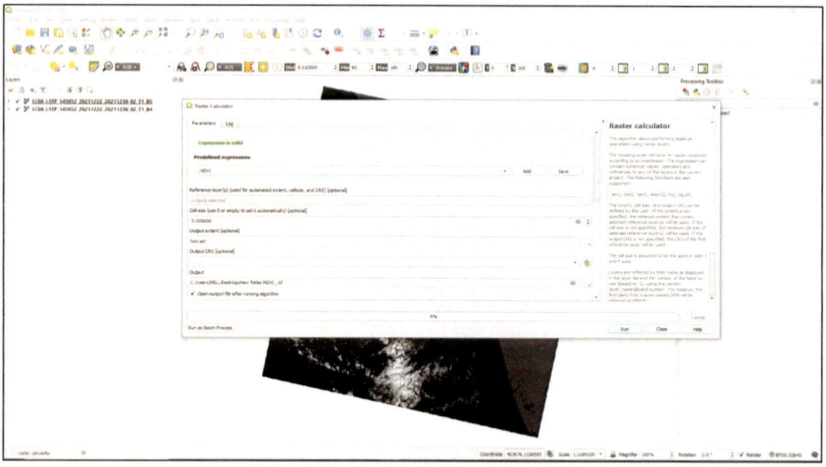
Raster calculator

4. NDVI image is generated. NDVI value ranges from −1 to 1.

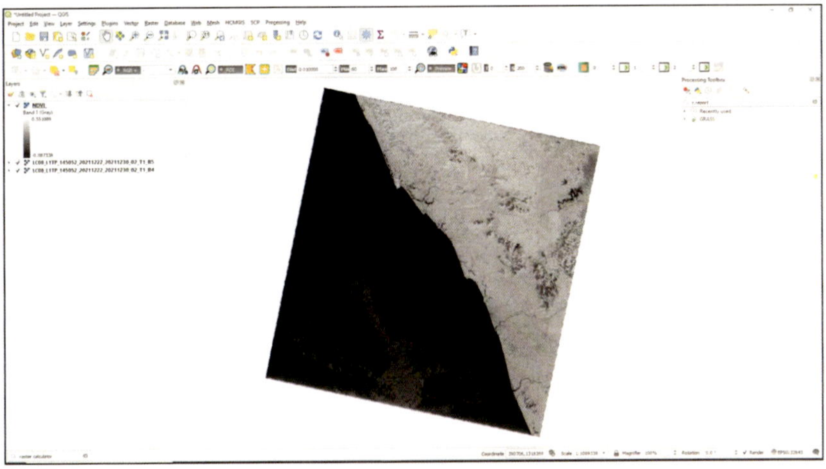

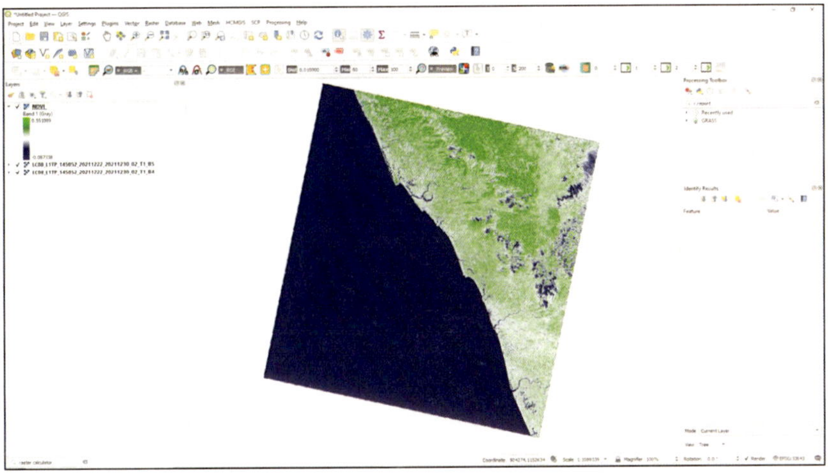

5. To calculate the area statistics, right click on NDVI image and go to *Symbology*. Change the mode to *Equal interval* and in Legend Settings, uncheck *"Use continuous legend"* and click *OK*.

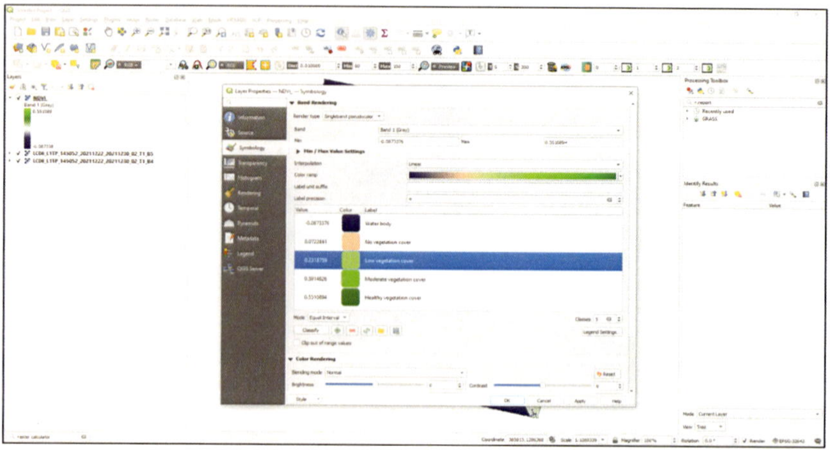

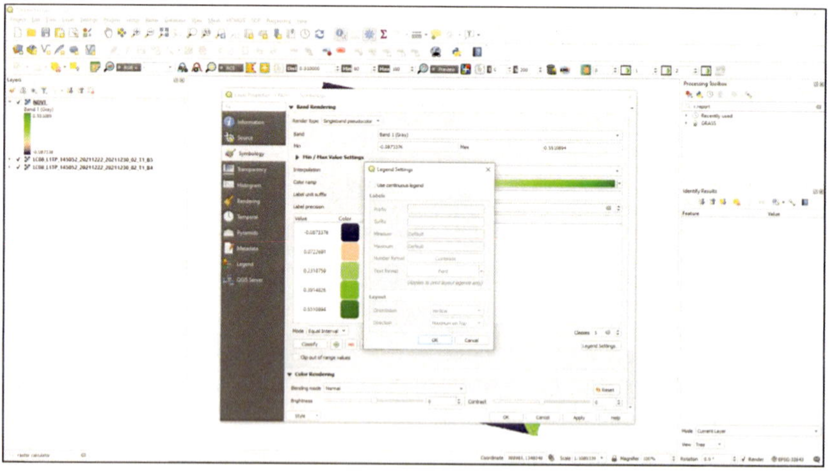

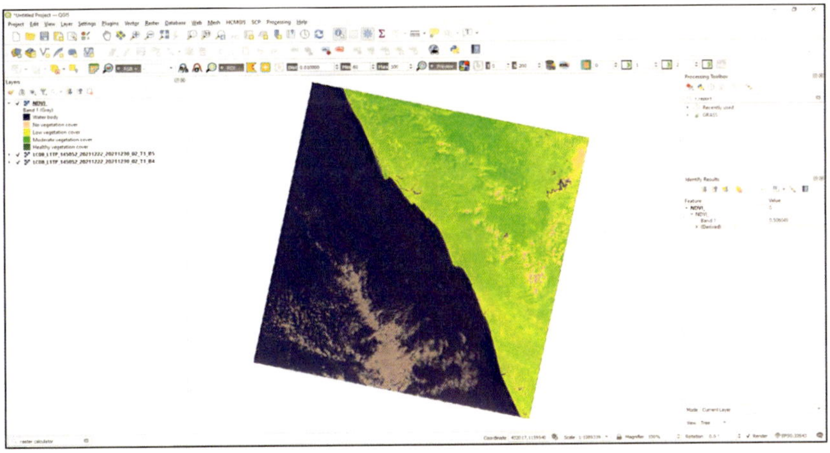

6. Go to *Processing* → *Toolbox*. Search for *Reclassify by table*. Fill the parameters and click *Run*.

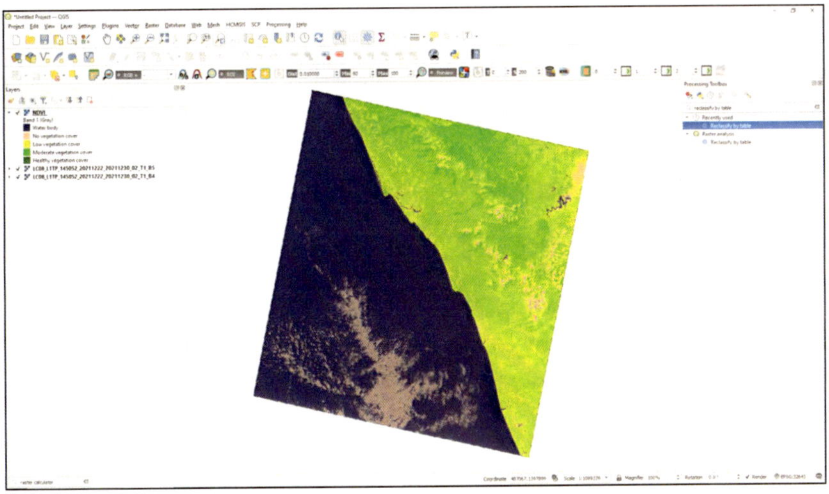

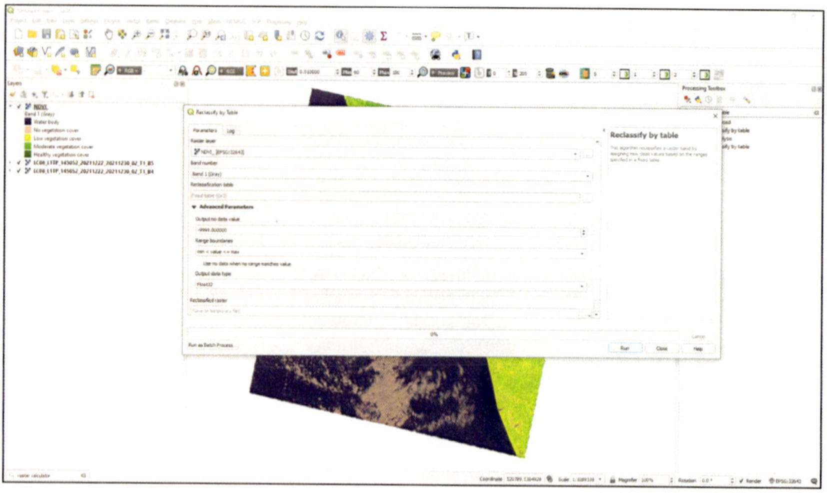
Reclassify by table

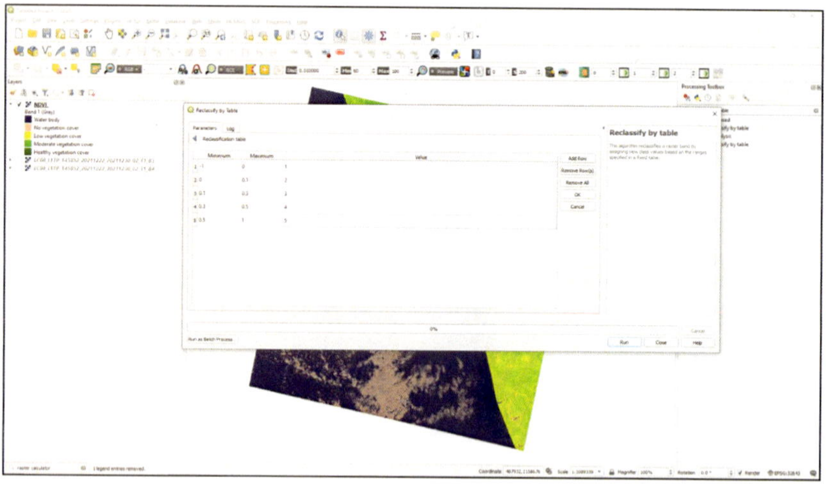
Reclassify by table

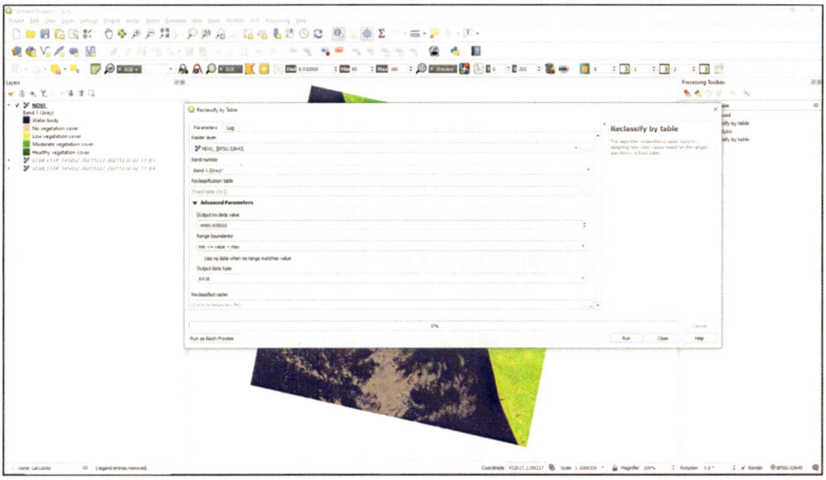
Reclassify by table

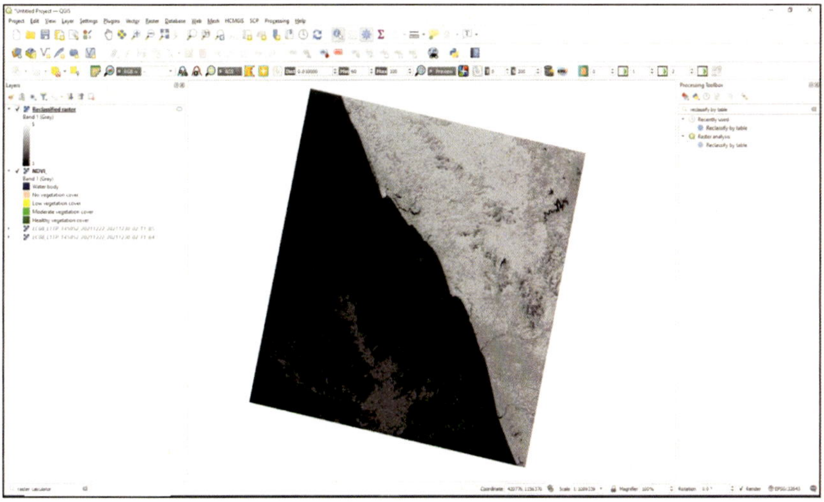

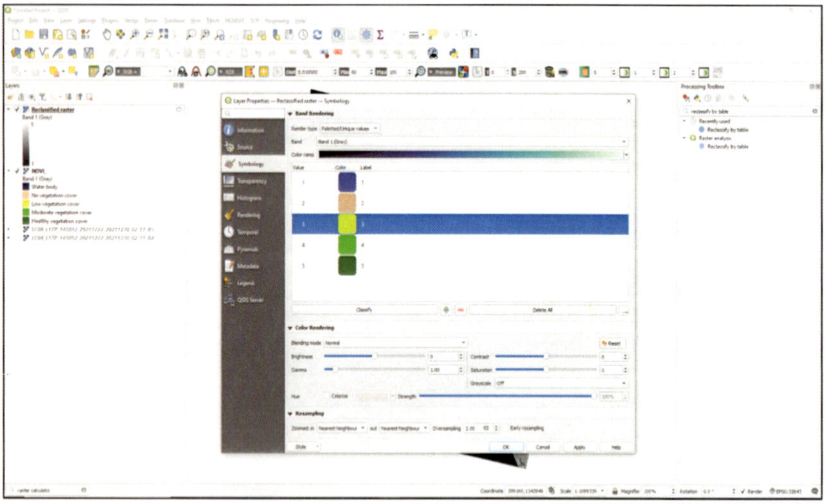

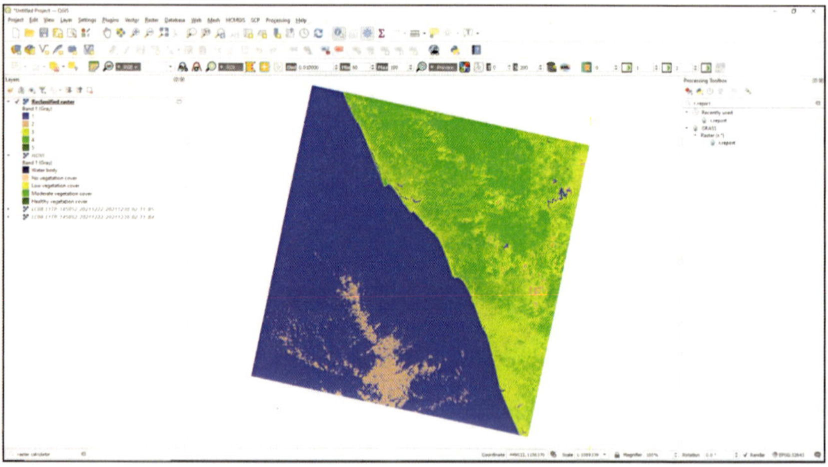

7. In *Processing Toolbox*, search for *"r.report"* and fill the parameters. Save the file as .txt.

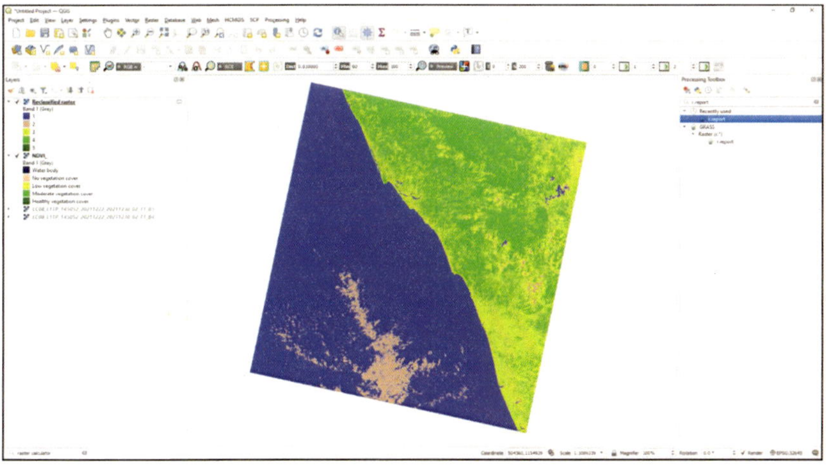

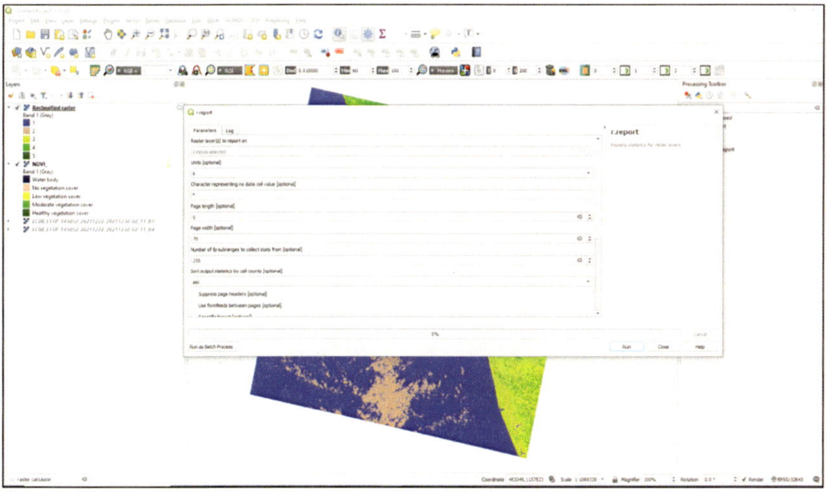

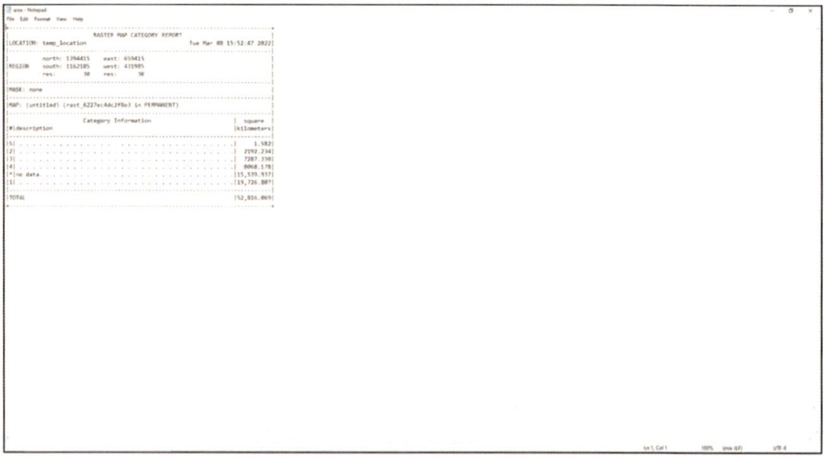

Supervised Classification

Theory

The process of extracting information classes from a multi-band raster image is referred to as image classification. The resulting layer can be used for creating thematic maps. Image classification can be further classified into two: supervised and unsupervised based on the interaction between the user and machine during the classification process.

Supervised classification uses the spectral signature defined in the training samples of each class and applies over to the whole image. Unsupervised classification identifies spectral classes from a multi-band raster image without the help of the user.

Procedure

1. Go to *Plugins* → *Manage and install plugins* → Search for *"SCP"* and click on *Install Plugin.*

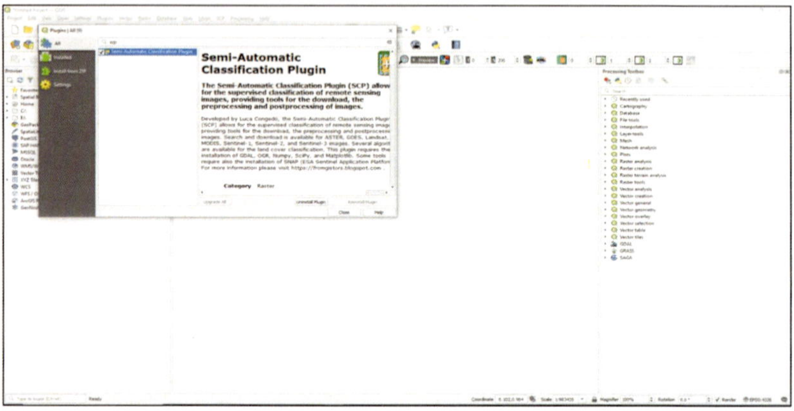

2. Go to *Layers Toolbar → Add layer → Add Raster Layer*. Select the required bands (in the case of Landsat 8, add bands 1 to 7).

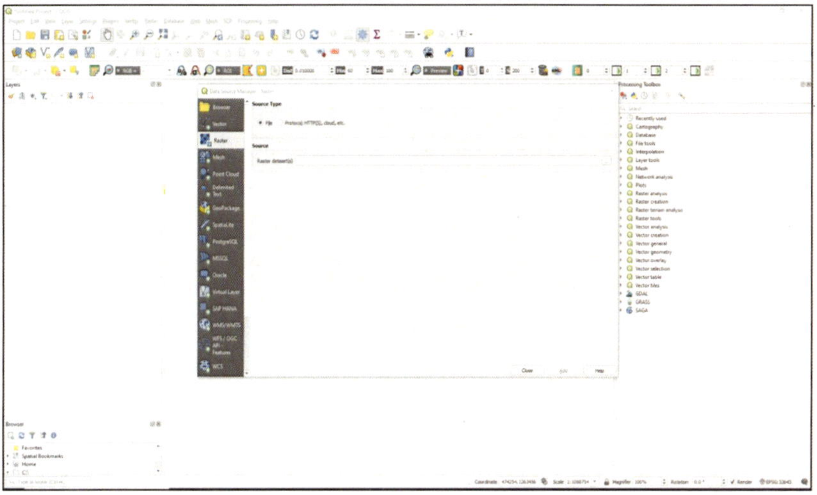

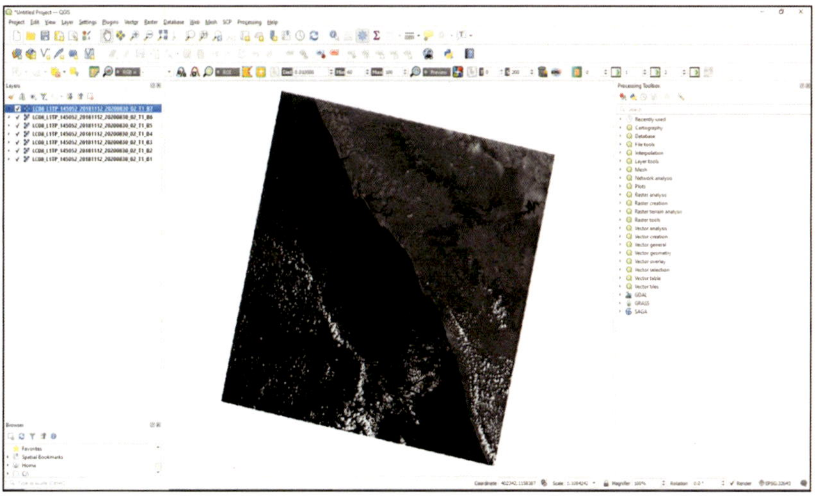

3. Click on *SCP → Bandset → Open a file →Select all the required bands*. The bands will be opened in *Single Bands List*. Select all the bands and click on *"Add band to band set"* [+] button.

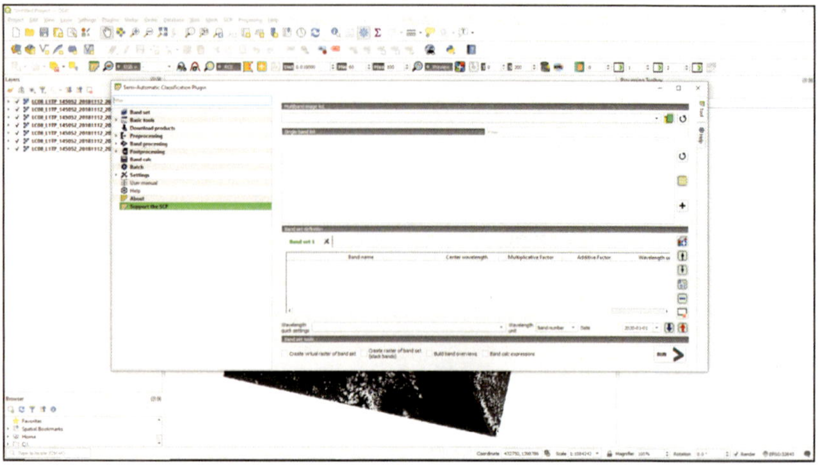

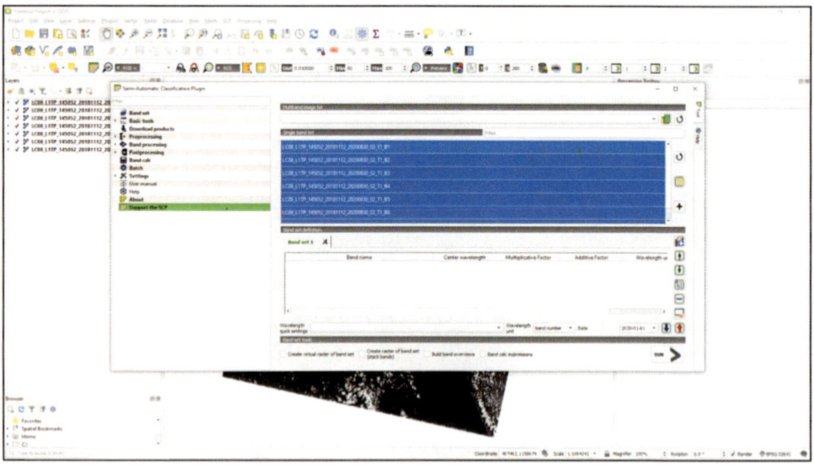

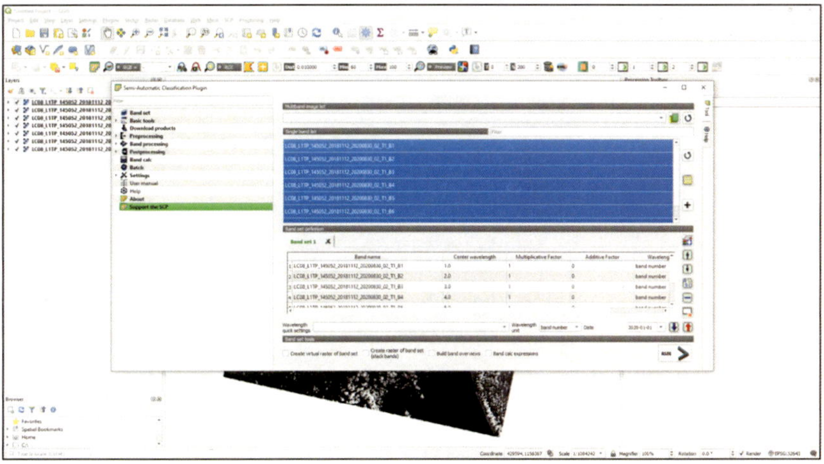

4. Go to *SCP* → *Preprocessing* → *Clip multiple rasters*. Provide the clip coordinates after clicking *Set area in the map* [+] button. Select the area and click *Run*. Create a new folder and select that folder.

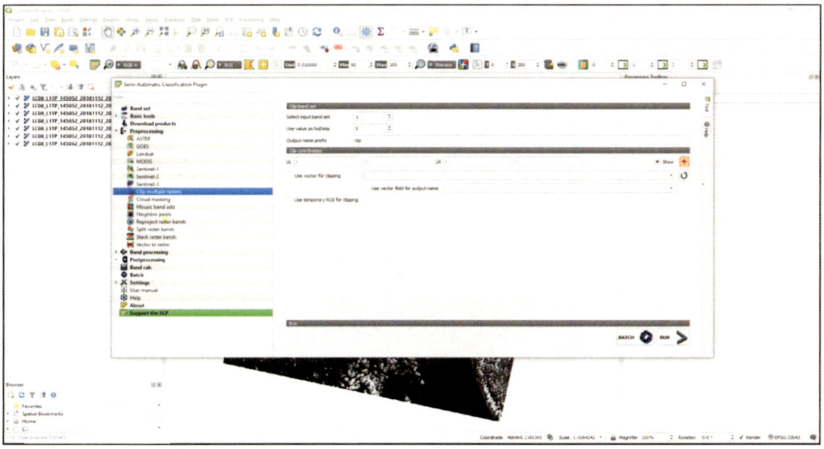

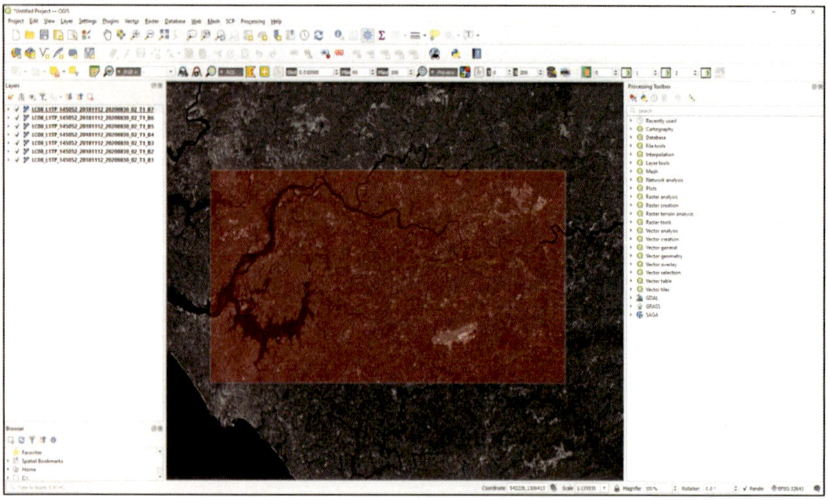

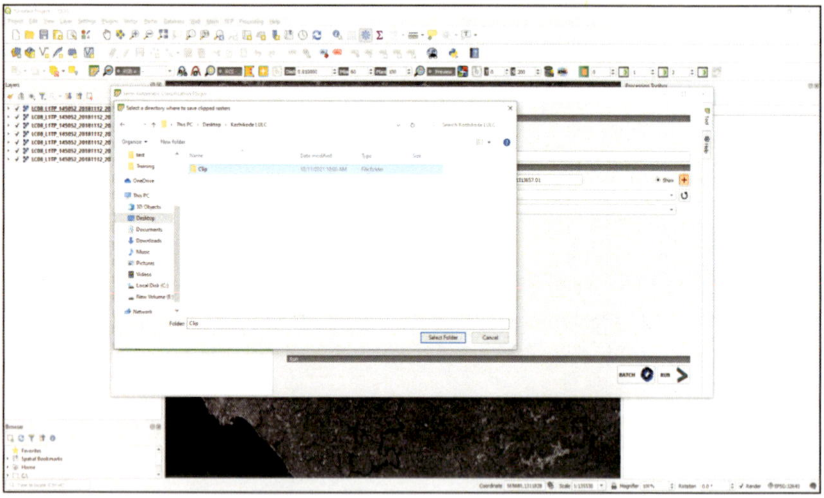

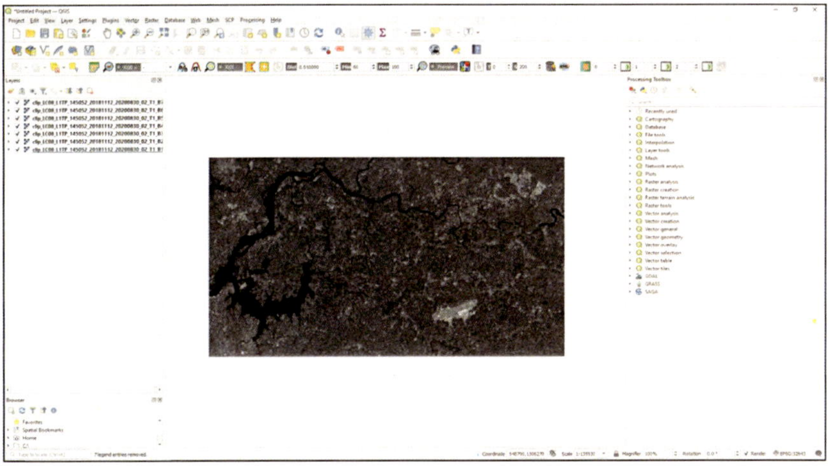

5. Go to *SCP → Bandset → Press Refresh* [] button so that the new set of bandsappears. Click the *reset button* [] and select the new bands and add the same to band set 1. Change the wavelength quick settings and the wavelength unit.

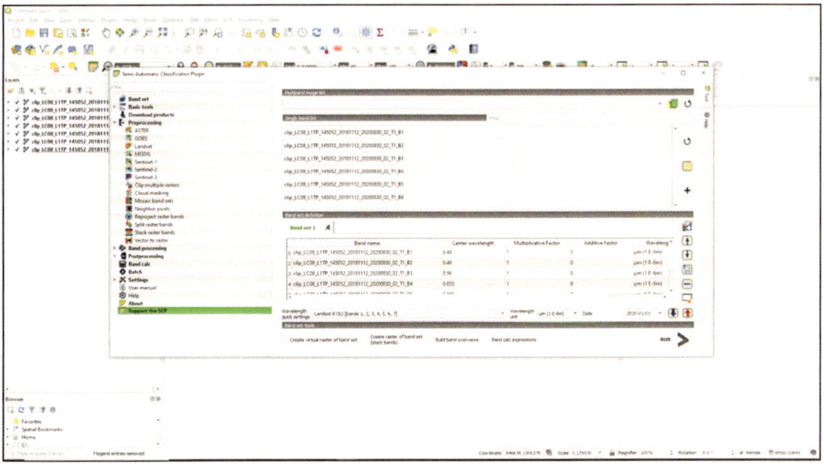

6. Go to *SCP* → *Preprocessing* → *Landsat*. Provide the parameters as shown in the figure below and click *Run*. Create a new folder and select it to process.

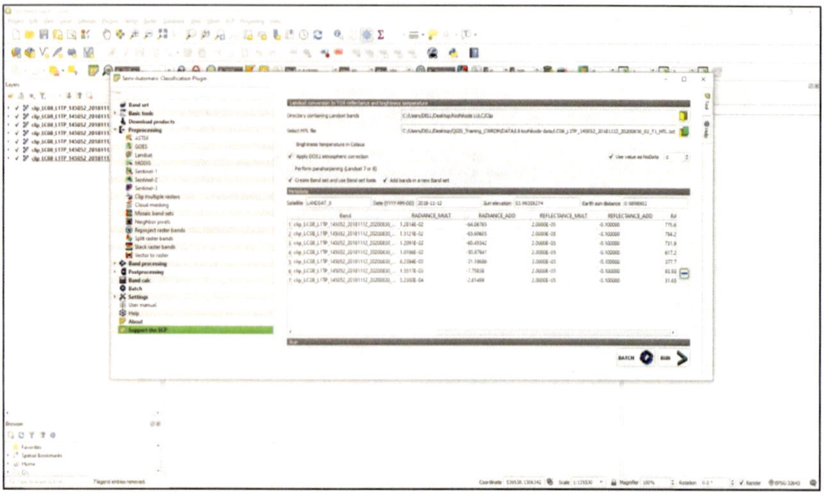

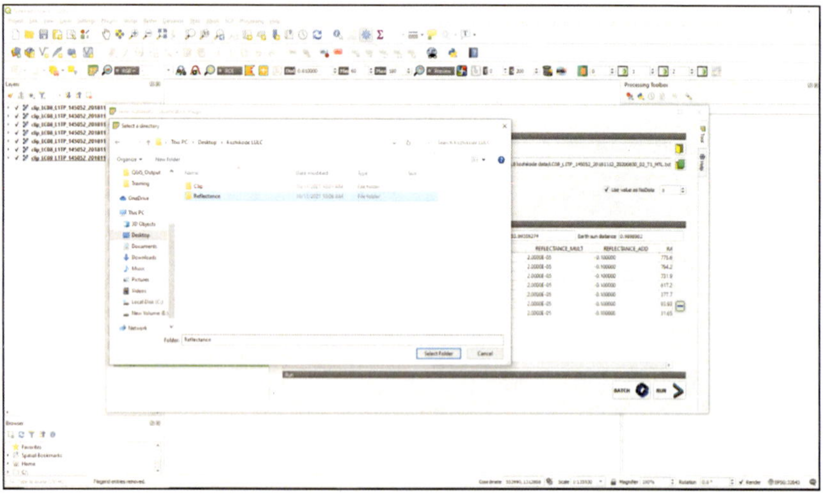

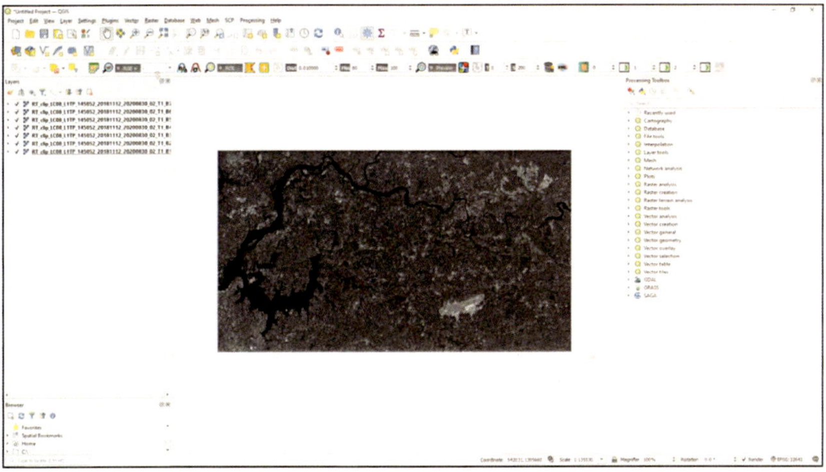

7. Go to *SCP* → *Bandset* → *Refresh the single band list*. Reset the band set and enter the new bands to band set 1.

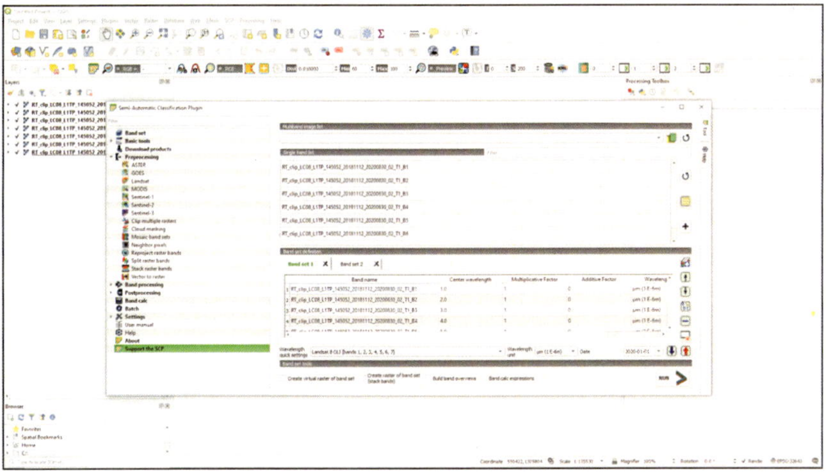

8. Minimize the tab and click on *R-G-B* [RGB =] (under SCP Working toolbar) and provide various combinations to see TCC, FCC, etc.

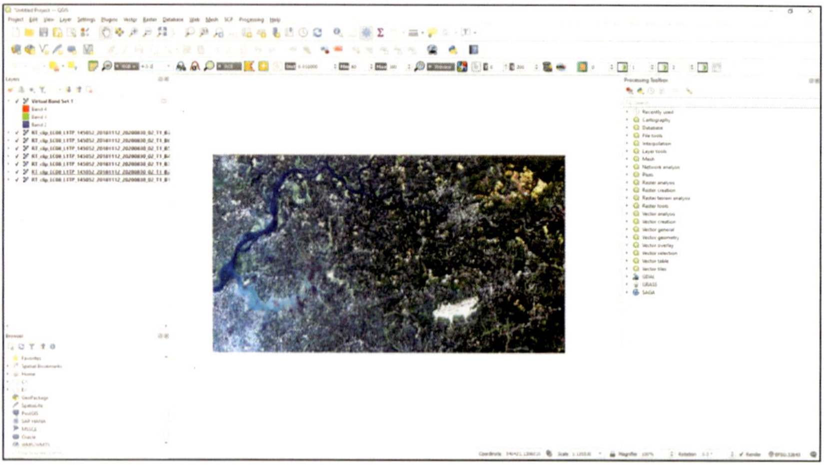

9. Click on *View* → *Panels* → *Enable SCP Dock*. Under SCP Dock, go to *Training input* and click on *Create a new training input* [ ]. Create a new folder and a file to save the training inputs.

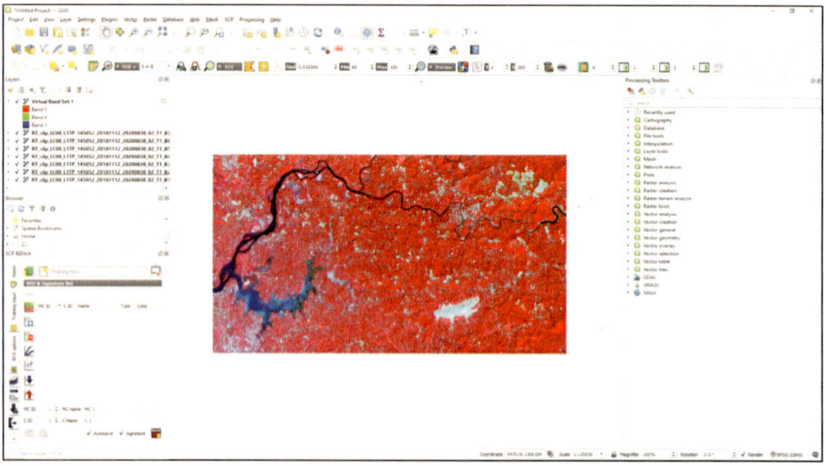

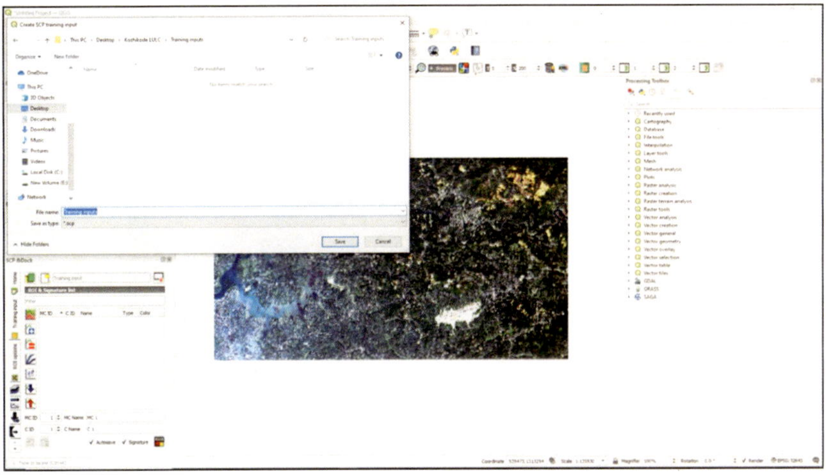

10. After creating a new macro class and class, click on *Create a ROI polygon* [ ] and create training classes for each land use and land cover. After selecting training inputs for each class, click on *save temporary ROI to training input* [ ].

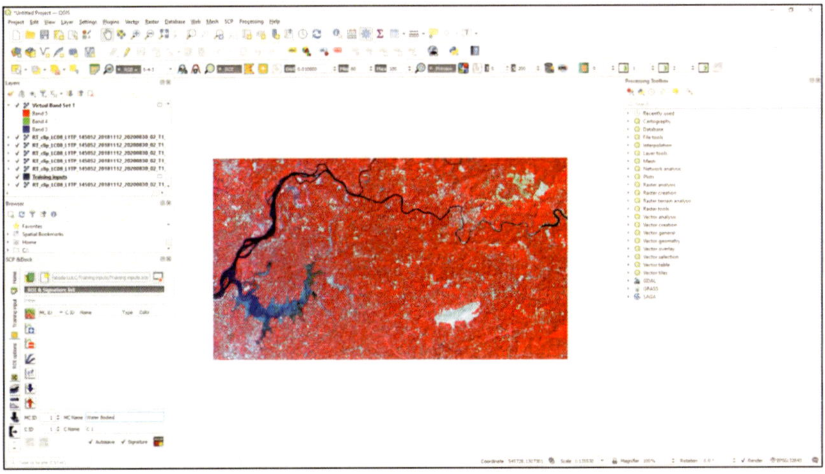

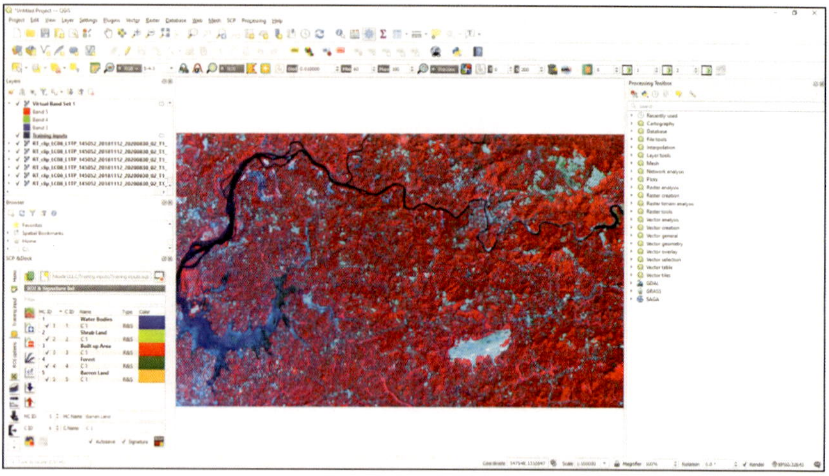

11. Go to *SCP* → *Band processing* → *Classification*. Minimize the tab and click on *Activate classification preview pointer* [ ] to get a preview of the classified image.

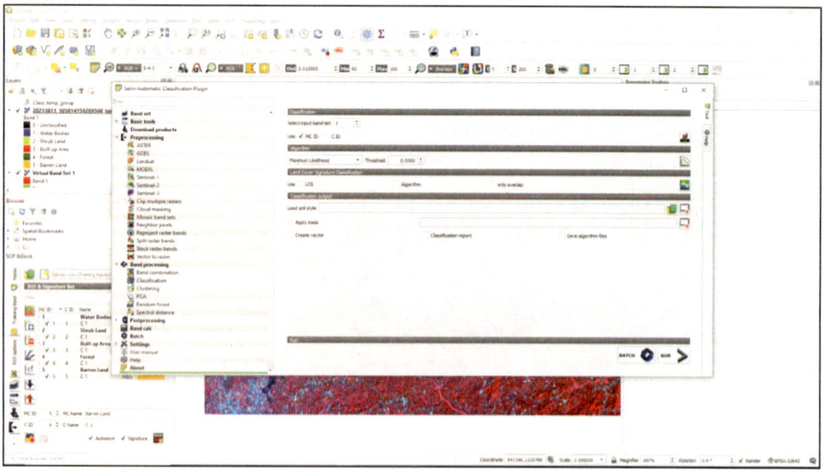

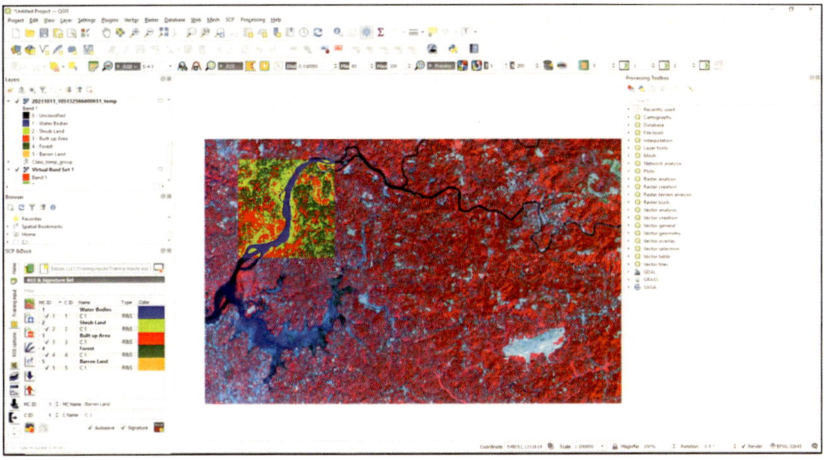

12. After viewing the preview, click on *Run* to get the classified image.

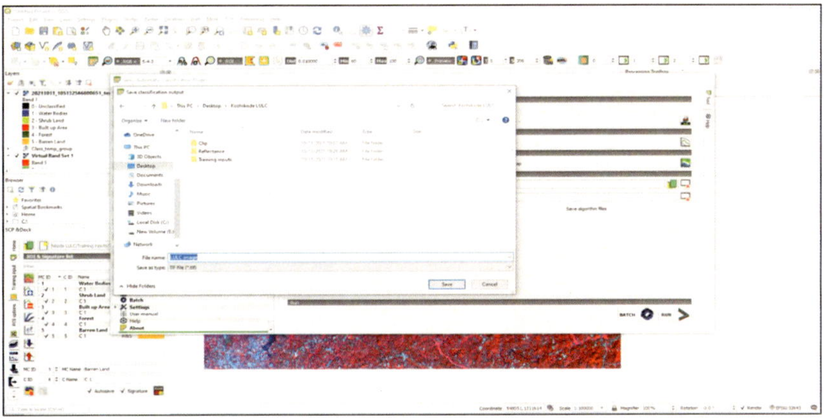

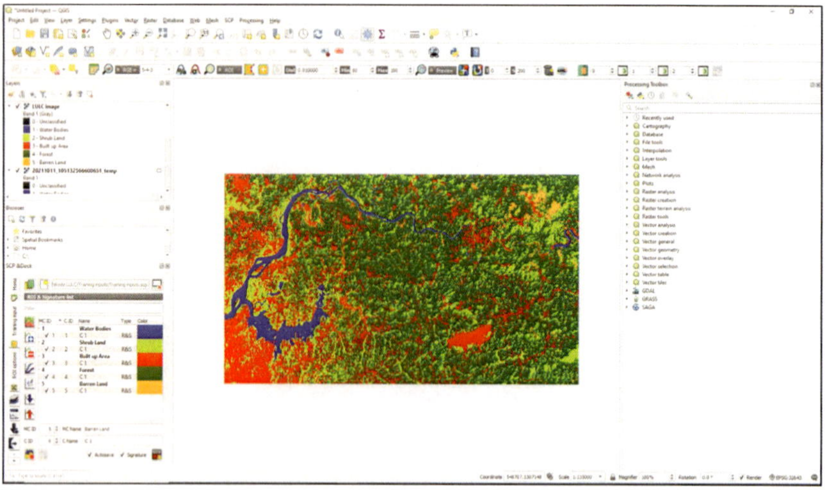

Watershed Delineation Using QGIS

1. Install QGIS 3.12. Go to *Layer* → *Add Layer* → *Add raster layer*. Browse and choose the DEM file.

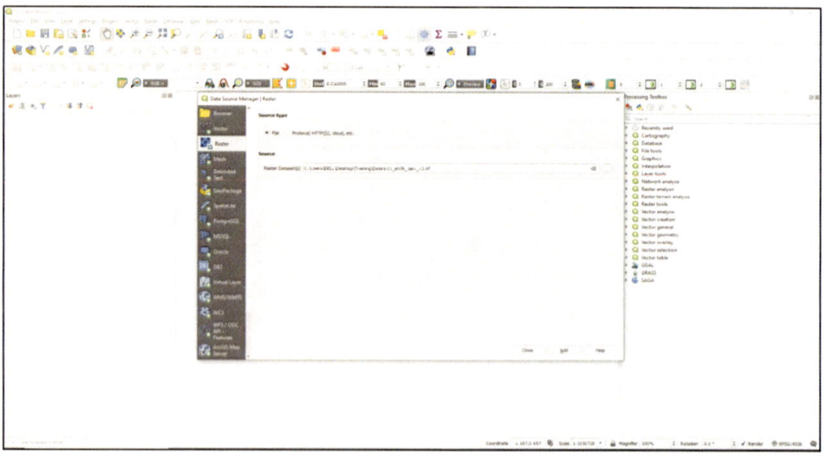

2. Go to *Raster* → *Projections* → *Warp (Reproject)*. Fill in the required parameters.

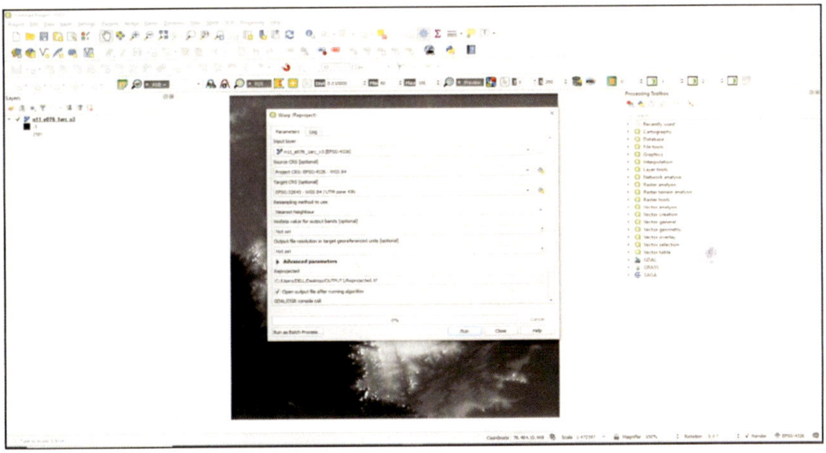

3. In the menu toolbar, click on Processing and enable toolbox. In the processing toolbox, search "fill sinks" and under SAGA Add-ons, select SAGA >*Terrain analysis-Hydrology* > *Fill sinks (wang & liu)*. Fill in the parameters.

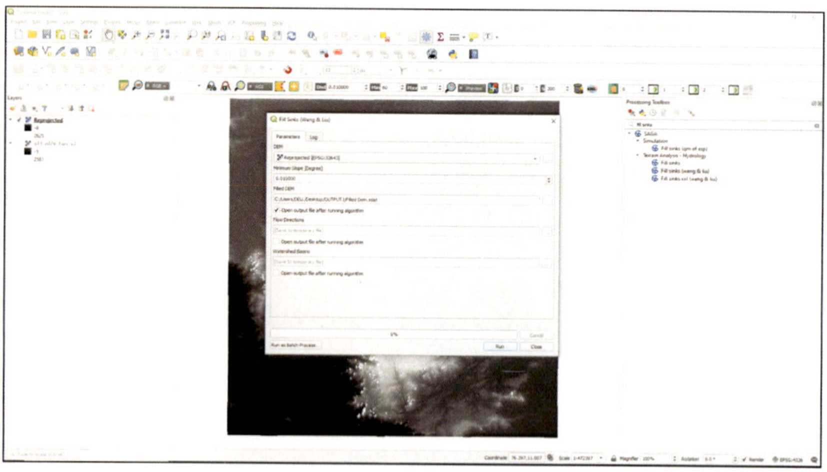

4. In the processing toolbox, search "Strahler order" and select *SAGA > Terrain Analysis-Channels > Strahler Order.*

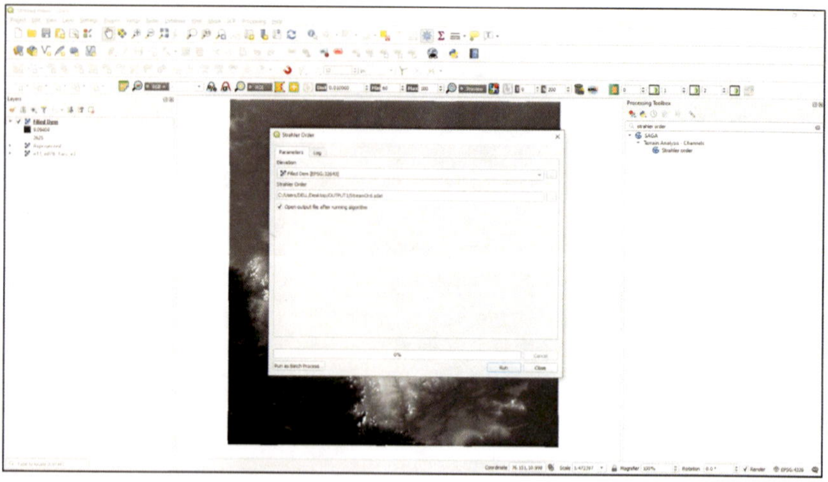

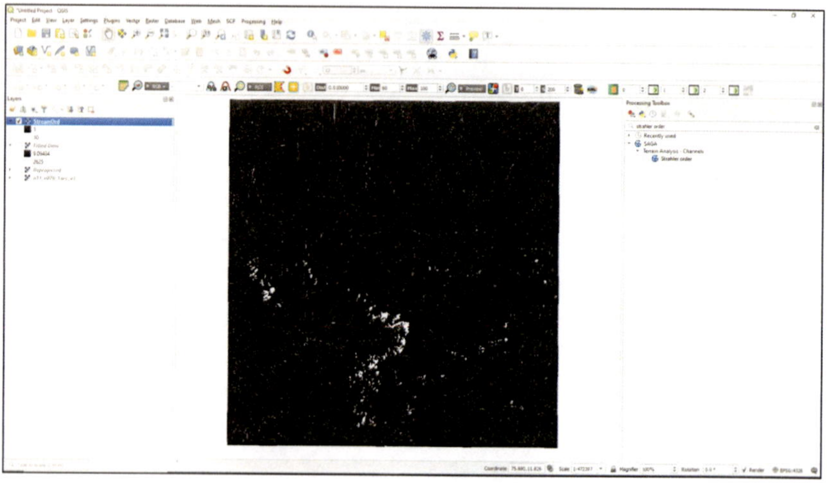

5. Go to *Raster* → *Raster calculator* and give the condition.

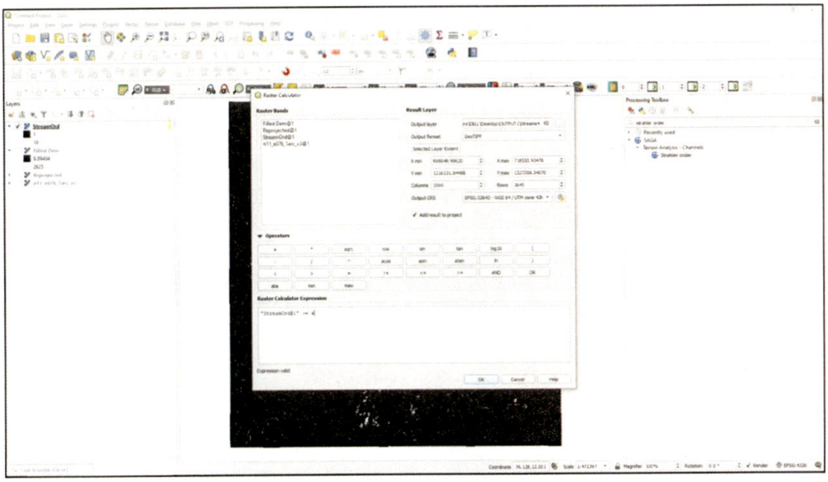

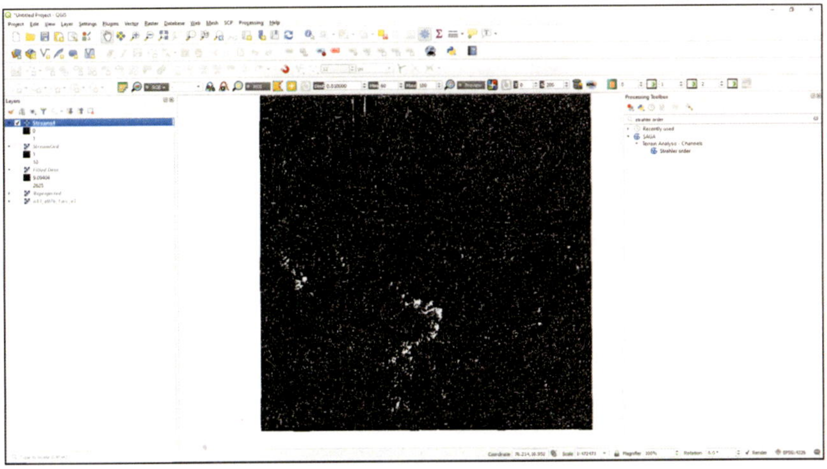

6. In the processing toolbox, search channel network and drainage basin and select *SAGA > Terrain analysis-Channels > Channel network and drainage basins.* Fill the required parameters.

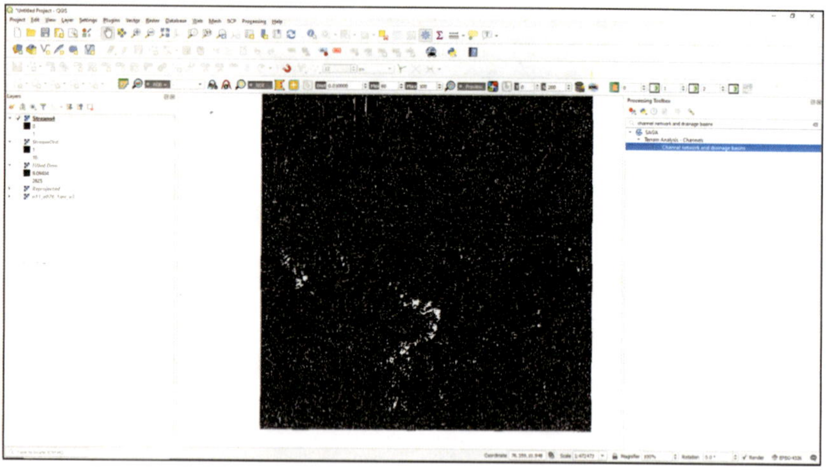

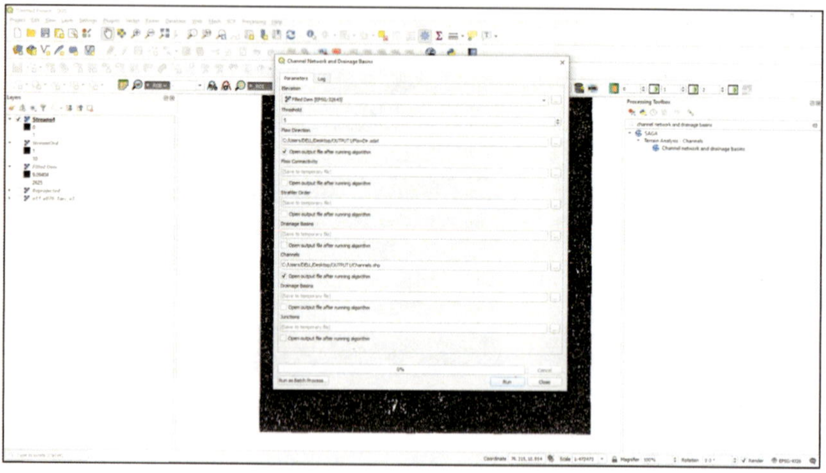

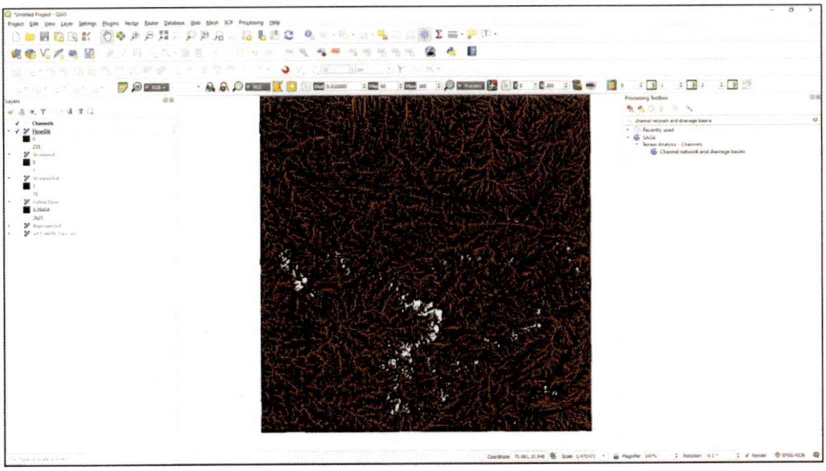

7. In the menu toolbar, click Plugins and search for Coordinate Capture. If not installed, install the plugin. If already installed, make sure the check box is ticked.

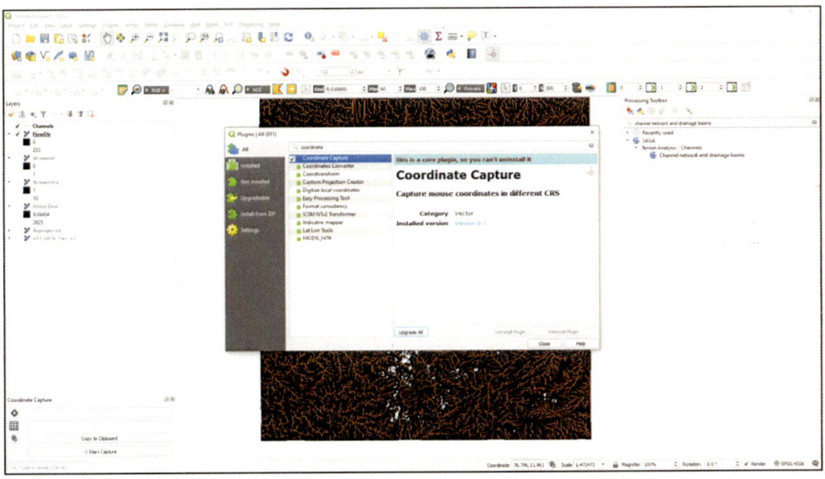

8. In the processing toolbox, search upslope area and select *SAGA > Terrain analysis- Hydrology > Upslope area.* Select the outlet point and with the help of Coordinate capture, note down the coordinates of the outlet point. (Use "Start Capture" tool under coordinate capture to note down the coordinates). Fill in other required parameters and run the algorithm.

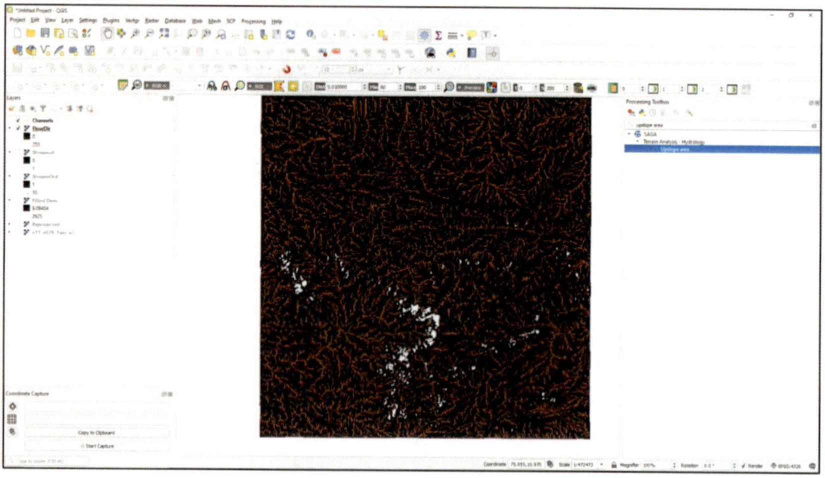

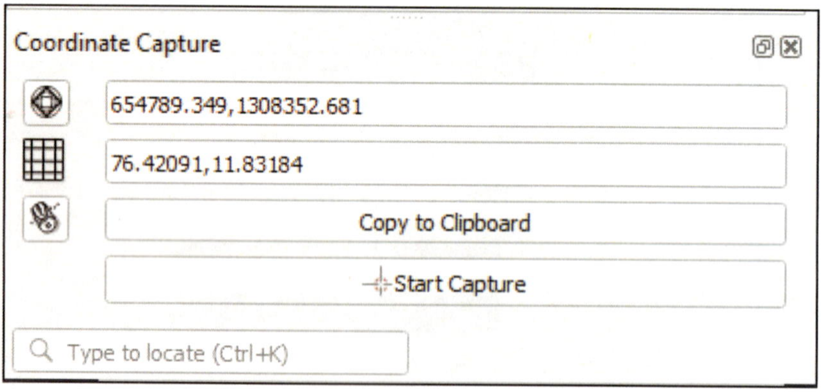

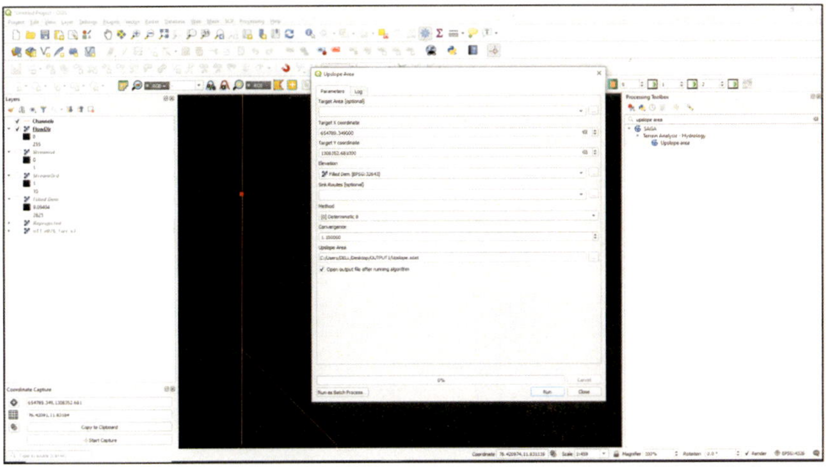

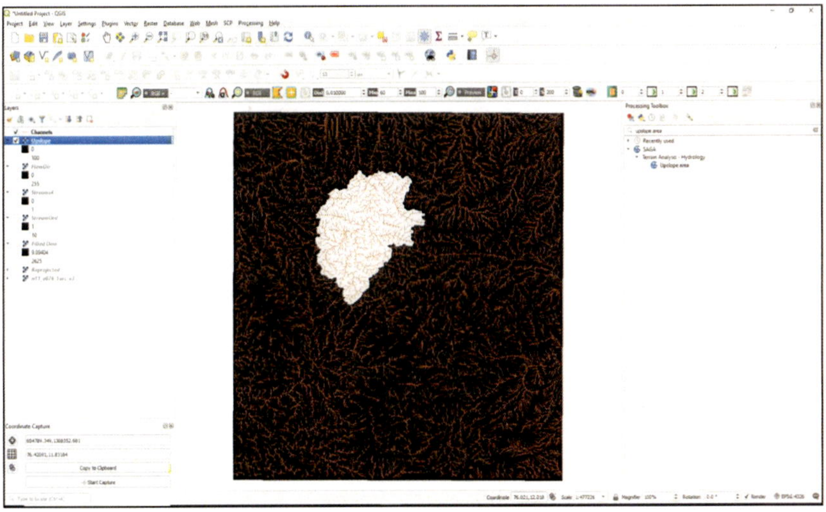

9. Go *to Raster → Conversion → Polygonize (Raster to vector)* and fill the required parameters.

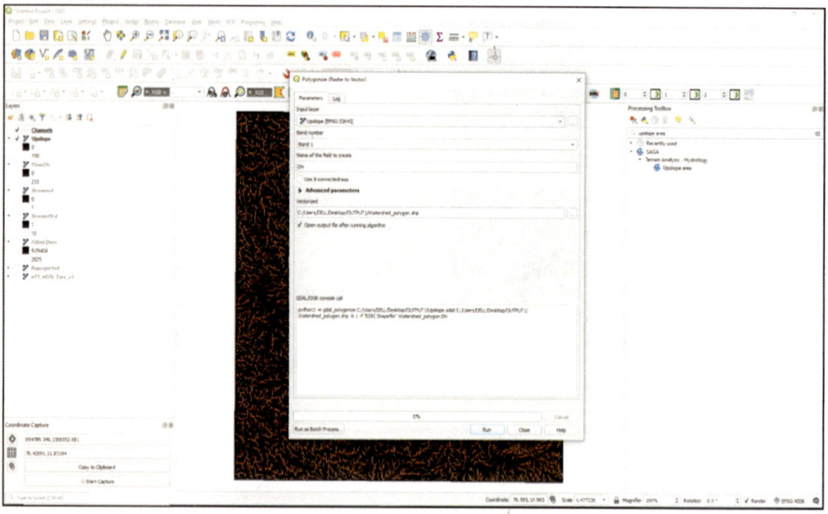

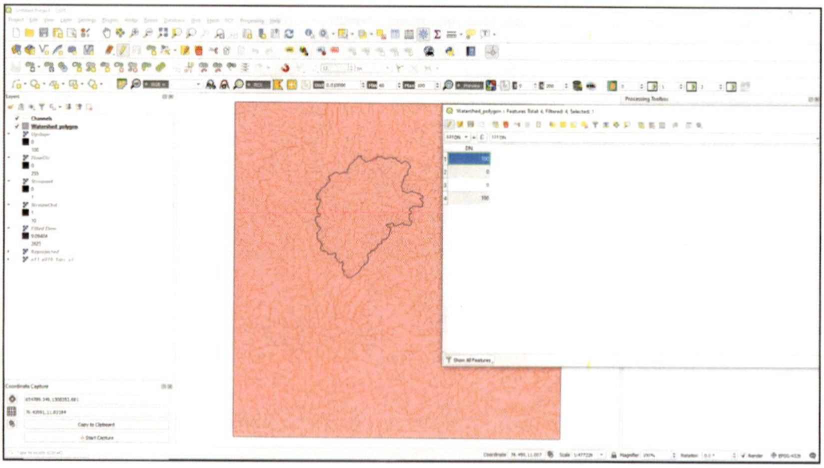

10. Right-click on the newly created shapefile of the watershed and select open attribute table. Click on *Toggle editing mode* [ ] anddelete the unwanted features using *Delete selected features* [ ]icon. Save the edits.

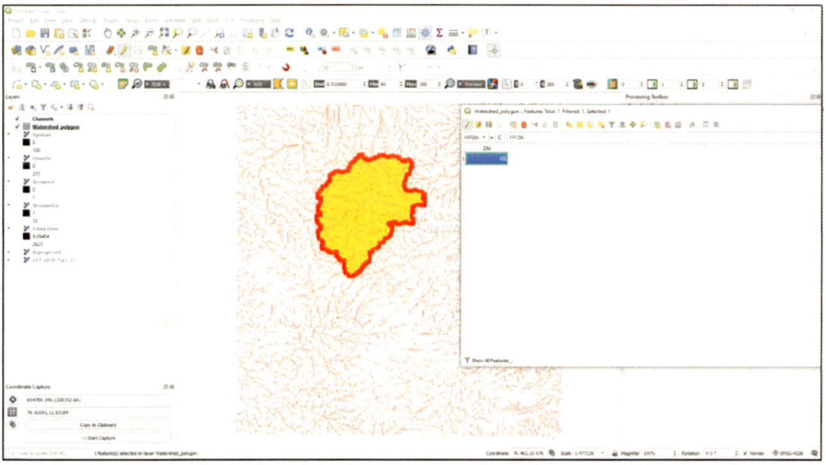

11. To calculate area of the watershed, open the attribute table of the shapefile. Click on *toggle editing mode* and click on *"New field."* After creating the field, click on *Open field calculator* to calculate the area of the watershed.

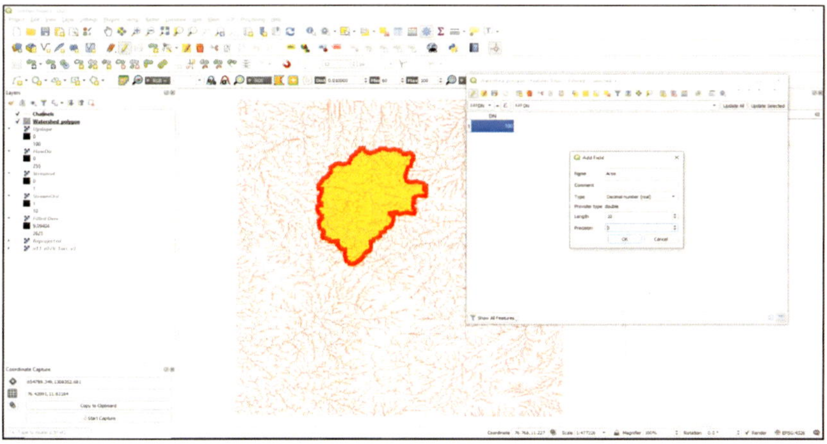

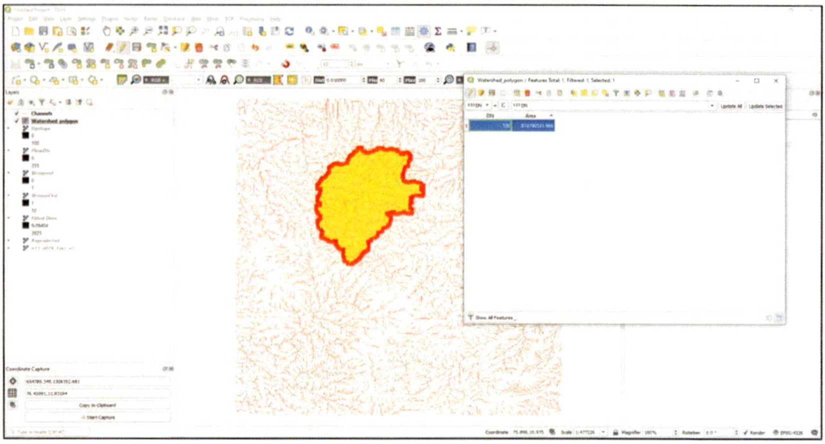

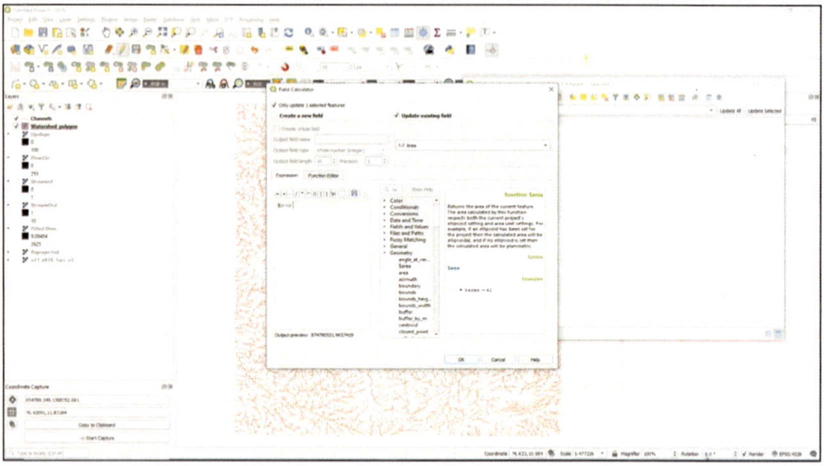

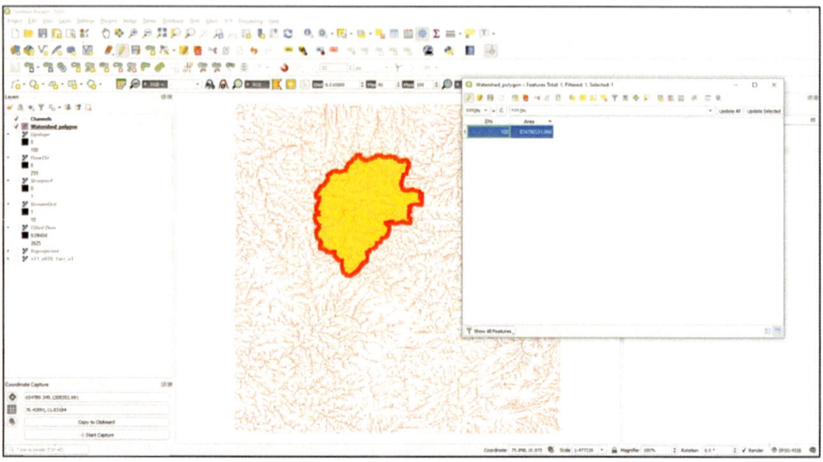

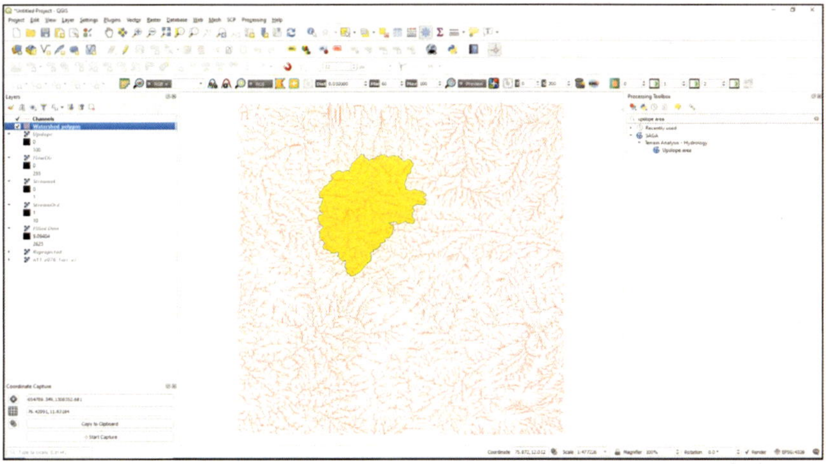

Acknowledgements: Authors are thankful to the open source QGIS platform (https://qgis.org/en/site/) under which the demonstration of GIS exercises are peformed, and thanks also to earth explorer (https://earthexplorer.usgs.gov/) for giving access to the freely available satellite remote sensing data for visualizing and performing the analysis in the GIS environment.

*Authors have used their own geodatabase for the demonstration of the GIS exercises, and the readers shall understand this and use their data accordingly to perform their analysis.